ROBES

0603574

REVENGE
OF THE
MICROBES

**How Bacterial Resistance
Is Undermining the
Antibiotic Miracle**

Abigail A. Salyers and Dixie D. Whitt
University of Illinois at Urbana-Champaign
Urbana, Illinois

**ASM
PRESS**

Washington, D.C.

Address editorial correspondence to ASM Press, 1752 N St. NW, Washington, DC 20036-2904, USA

Send orders to ASM Press, P.O. Box 605, Herndon, VA 20172, USA
Phone: (800) 546-2416 or (703) 661-1593
Fax: (703) 661-1501
E-mail: books@asmusa.org
Online: www.asmpress.org

Library of Congress Cataloging-in-Publication Data

Salyers, Abigail A.
  Revenge of the microbes : how bacterial resistance is undermining the antibiotic miracle / Abigail A. Salyers and Dixie D. Whitt.
      p. ; cm.
  Includes bibliographical references and index.
  ISBN 1-55581-298-8
  1. Drug resistance in microorganisms—Popular works.
  [DNLM: 1. Drug Resistance, Microbial—Popular Works. QW 52 S186r 2005] I. Whitt, Dixie D. II. Title.

  QR177.S26 2005
  616.9'201061—dc22

                                                                    2005002847

10 9 8 7 6 5 4 3 2 1

Cover photo: Scanning electron micrograph of *Staphylococcus epidermidis*.
© Dr. Dennis Kunkel/Visuals Unlimited.

For Jeff and Greg

Contents

Preface

Most people have a love-hate relationship with antibiotics. They love the fact that antibiotics still work most of the time and work quickly with few side effects. Anyone who has experienced therapy that only suppresses the symptoms of a disease, such as arthritis medications, or has endured therapy that is debilitating, such as cancer chemotherapy, has to appreciate the swift, effective action of antibiotics. But people also hate things about antibiotics. They hate the fact that more and more physicians are refusing to prescribe antibiotics for flu and that patients are being blamed for demanding antibiotics when they shouldn't. They are also coming to hate the dire warnings about increasingly resistant bacteria whose advance may send us back to the preantibiotic era.

Many people, including some scientists, are also confused about antibiotics. Why can't I have antibiotics when I have a bad case of the flu? I'm sick, aren't I? Why do research scientists and public health officials tell me that I should be worried about bacteria that have become resistant to antibiotics while many physicians still deny that bacterial resistance to antibiotics is a significant clinical problem? What are antibiotics, anyway? And what does antibiotic resistance mean? Does it mean that I become resistant to antibiotics or that the bacteria do? Should I worry about the use of antibiotics in agriculture or should I listen to the spokespeople for the farmers' organizations who reassure me that all is well?

It was this ambivalence and confusion that motivated us to come out of our comfortable niche as writers of textbooks and attempt to write a book for the general public about antibiotics and resistance to them. In undertaking this challenge, we began to realize some things that surprised us. First, college students receive virtually no information about antibiotics or other antimicrobial compounds. Even medical students in most medical schools get precious little instruction until they take pharmacology in their later years. We have a whole generation of physicians, now in their 50s, whose education in microbiology may have consisted of only two or three weeks of lectures, with antimicrobial compounds taking up only a small slice of that time period. Perhaps even more surprising, most graduate students in biology receive little or no education in the field of antibiotics and antibiotic resistance. In fact, the one thing most of them learn about antibiotics and antibiotic resistance is that these are old fashioned topics that no one is interested in anymore.

The second thing that surprised us is the extent to which average people are interested in the topic. We had never underestimated the intelligence of people in general, once their attention has been engaged, but we had doubts about the level of their interest. That doubt changed during the period after the anthrax attacks in October 2001, when we and some of our colleagues spent hours talking to postal workers, university staff members, and undergraduates in introductory classes. They wanted to know more than the answer to the question, what options do I have? They wanted to know what bacteria are, how antibiotics work, and why government and media workers were getting one antibiotic while postal workers were getting another. These questions did not end once the anthrax panic had subsided. They had already shown up in concerns about the safety of genetically modified plants, which contain bacterial antbiotic resistance genes, or, more recently, about the consequences of the use of antibiotics in agriculture. Granted, those who were most immediately affected by the anthrax attack or by the debate over the agricultural use of antibiotics had a strong vested interest in paying attention. But the interest went beyond that. People want to know what is going on with their health and with the health of the community.

Something we relearned, because in a way we already knew it, is that a book about antibiotics and antibiotic resistance has to go beyond the scientific facts and try to deal with the social, economic, and political aspects of the topic. Scientists like us are not used to writing about such things. It's a little like explaining to your grandmother, who has asked about condoms, how condoms give partial protection against sexually transmitted diseases. You feel a little embarrassed, not just because of the intimate nature of the topic but because you are not sure you are competent to convey all the subtleties of the topic. Scientists are not trained formally to speak about such things as social, economic, and political aspects of scientific advances. We received this training in a different milieu. In a sense, we both have earned second Ph.D.s in dealing with these subjects in the school of hard knocks: teaching, testifying before regulatory agencies, talking to reporters, and talking to community groups. Of course we didn't get it right every time, but we didn't flunk either.

A different kind of challenge we faced with this book was to decide who the audience was. Every other book on antibiotics and antibiotic resistance that has stressed the scientific aspects of the topic has had scientists with advanced degrees as its audience. We wanted to reach a wider audience, but we didn't want to exclude scientists, especially those who work in areas in which information about antibiotics might not have been part of their training.

Accordingly, we made some compromises. For example, we decided that we wanted to show the chemical structures of antibiotics because some people would be interested in seeing them. However, they tended to break up the narrative flow of the text and are somewhat off-putting to those who are not used to seeing them, so we did what all scholarly types do when faced with such a dilemma: we put them in an appendix. We also decided not to list references to scientific papers at the ends of the chapters. We put those in an appendix too. Do you begin to detect a pattern here? We are so pathetically predictable.

We also did something else that is usually not seen in books like this one. We end each chapter with some questions labeled "issues to ponder." Normally, we are only too happy to ram our opinions down the throats of those who can't run faster than we can. (Fortunately, our advancing age has made this less of a threat than it once was.) However, we wanted to convey the very important message that people, whether they have scientific credentials or not, should be able to have and express opinions about controversies in the areas covered in the different chapters. We realize that this device labels us as a couple of unreconstructed fusty old pedants, but it was the best thing we could come up with.

Finally, we did not want to follow the example of others who have written about this issue and forecast gloom and doom or spread blame and pain, because we want readers to enjoy this book. Admittedly, it is a book about a serious subject, but there are light as well as dark sides to the subjects it covers. There are three things we hope readers will take away from this book. First, there is plenty of hope. There is still time to avert what is constantly being portrayed as the impending disaster of a return to the preantibiotic world. Second, everyone should and can have a say in what is done about saving the antibiotics we already have and battling antibiotic-resistant bacteria. However, to be part of the solution, you need to have information. We hope this book provides this information in a relatively painless way. Third (and this is probably a wildly unrealistic expectation), we hope that this book leaves people who have not thought much about bacteria with an appreciation for these indomitable little critters. This may sound like an odd goal, attached as it is to a book about the damage bacteria can do, but the important message we hope this book brings is that these tiny parasites are not consciously malevolent. They were here long before we appeared, and in a very real sense, they gave us the possibility of life. When they cause us problems, remember that they're just trying to make a living. And they're not even making minimum wage.

Abigail Salyers
Dixie Whitt
January 2005

1

Magic Bullets, Miracle Drugs

When antibiotics were first introduced, they were regarded as miraculous compounds. A dubious mark of medical progress is that most people today have lost this earlier awe of antibiotics. Sadly, we have come to take antibiotics for granted. In fact, we take antibiotics so much for granted that many of us willingly participate in the misuse and overuse of antibiotics without thinking of the potentially disastrous consequences of our actions.

An often-cited example of antibiotic misuse is the use of antibiotics, which are antibacterial compounds, in an attempt to combat a viral infection such as the common cold or the flu. Physicians know that using antibiotics for this purpose is not good medical practice, but many yield to pressure from patients who are too sick and too concerned with job and family pressures to care much about whether an antibiotic is deemed by the experts to be appropriate for their condition. In the mind of the patient, the antibiotic can't hurt and it might help.

Misuse of antibiotics has consequences. It may not cause immediate damage to the patient being treated, but it is dangerous for the population as a whole. Overuse of antibiotics, for whatever purpose, leads to a reduction in the effectiveness of the antibiotic by increasing the selection for bacteria that have become resistant to it. The more antibiotics are used, the more bacteria that are resistant to them predominate. Clearly, we do not want to stop antibiotic use entirely, but using antibiotics responsibly, only when their use is justified, is in the public interest.

Shaking an admonitory finger at physicians and patients who misuse antibiotics may make the finger shaker feel good, but face facts: this strategy is not working very well. Only when people have the information on which to base informed decisions about their own health care will it be possible to gain wide acceptance of the need to scrutinize and limit antibiotic use to those cases in which it is justified. This book is an attempt

1

to provide such information, along with a view of the issues relating to antibiotic use that are currently being debated in the medical community.

What Are Antibiotics?

We all know that antibiotics are good for us. Otherwise, why would there be so much pressure on physicians to overuse them? But what are antibiotics? The short answer to this question is that antibiotics are chemical compounds that kill or inhibit the growth of bacteria. This answer is somewhat misleading, however. Many compounds such as arsenic or mercury can kill bacteria, but few people would volunteer to take large doses of arsenic orally or by injection. What distinguishes antibiotics from substances such as arsenic and mercury is that antibiotics act in ways that are specific for bacteria, not as general toxins that harm both the bacteria and the human body.

The revolutionary feature of antibiotics, compared to earlier therapies for infectious diseases, is that they are designed to be devastating for bacteria but to have few if any adverse effects on the human body. This is the reason antibiotics have been called "magic bullets," compounds that target invading bacteria and leave the human body alone. Of course, some antibiotics do have adverse side effects in some people. Allergy to penicillin is a well-known example. In general, though, antibiotics are gentle cures.

Most antibiotics are made by bacteria or fungi, but an increasing number are either synthesized by chemists (synthetic antibiotics) or are chemically modified in the laboratory to increase their potency (semisynthetic antibiotics). Older scientists still make a distinction between antibiotics (with *biotic* referring to their biological origin) and synthetic antibacterial compounds. They rightly prefer the term *antibacterials*, a term that embraces both types of compound. But as so often happens, such distinctions blur with time and use, and many scientists today use the terms *antibacterial* and *antibiotic* interchangeably. We will use *antibiotic* instead of *antibacterial*, especially in view of the fact that most nonscientists are more familiar with the term *antibiotic*.

How Do Antibiotics Differ from Disinfectants and Vaccines?

When we were talking to people in preparation for writing this book, we found that in addition to wanting to know what antibiotics were, people also wanted to know how antibiotics differed from other antimicrobial compounds such as disinfectants and antiseptics and how they differed from vaccines. Although this book focuses on antibiotics, it is worth addressing this question because the most effective strategy against bacterial infections is a strategy that combines these different anti-infective approaches.

Antiseptics are compounds that are too toxic for internal use but can be safely applied to skin. Examples are triclosan (a common ingredient in antibacterial chopping boards, toys, and soaps), peroxide, and mercury derivatives such as mercurochrome. Disinfectants are compounds that are too toxic to be used to cleanse skin but are valuable agents for cleaning inanimate surfaces. These include household bleach and formaldehyde.

Disinfectants and antiseptics prevent infection by reducing the number of bacteria that have access to vulnerable areas like cuts or surgical wounds. They are nonspecific in their killing power. In fact, many kill viruses and fungi as well as bacteria. Antibiotics, which are designed to be specific for bacteria, can be administered preventively, a process called antibiotic prophylaxis. For example, patients who are receiving aggressive cancer chemotherapy that makes them temporarily at high risk for developing a bacterial infection are often given antibiotics to prevent an infection from developing. Antibiotics, however, are mainly used in human medicine to treat an infection that is already under way.

A different strategy for preventing infectious diseases is vaccination. Vaccines prevent disease by priming the human immune system to attack a certain type of infectious microbe. Vaccines prepare immune cells to respond to molecules on a microbe's surface, such as specific proteins or carbohydrates, so that the immune cells leap into action within hours after a microbe invades.

Ordinarily, weeks elapse between the first contact with a specific bacterium or virus and the development of an effective immune response. In other words, vaccination speeds up this reaction period by giving the appropriate immune cells a "dress rehearsal" in which the threat confronted by the immune system is a nontoxic mimic of the actual infecting microbe. Vaccination is specific for a particular microbe, whereas antibiotics attack many different kinds of bacteria.

All of these layers of protection are important bulwarks against infectious disease. They are most effective when used in combination to decrease a microbe's opportunity to invade and infect the human body. If the preventive measures fail, antibiotics are there to intervene as a second line of defense.

The "Miracle" in "Miracle Drugs"

When antibiotics first appeared in the 1930s and 1940s, people held them in much greater awe than we do today. To gain an appreciation for why this was so, let's take a time trip back to the bad old days before antibiotics were available and consider some examples of the impact of antibiotics on medical practice. We will start with some old infectious disease scourges and work our way into the present.

Syphilis

If you were a sexually active man or woman in a pre-20th century U.S. or European city, and especially if you consorted with prostitutes, syphilis was a very real threat. Today we know that syphilis is caused by a corkscrew-shaped bacterium, *Treponema pallidum* (Fig. 1.1).

Untreated syphilis progresses through three phases. In the initial phase, growth of the bacteria in the site of inoculation produces an ugly

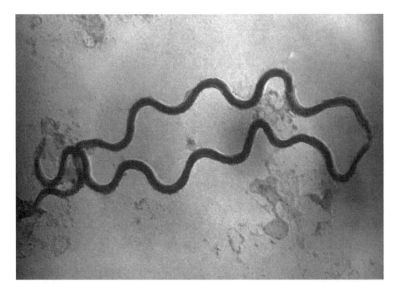

Figure 1.1   *Treponema pallidum*, the causative agent of syphilis, as seen with an electron microscope. (Courtesy of Janice Carr, CDC, Atlanta, Ga.)

inflamed sore called a chancre. Although the chancre looks awful, it is seldom painful. The chancre heals spontaneously after a week or two, giving the infected person a false sense of security. In the secondary phase of the disease, the bacteria enter the bloodstream. At this point, a rash may appear, especially on the palms of the hands or the soles of the feet. As in the case of the chancre, the rash disappears spontaneously, even though the bacteria are still present.

T. pallidum then moves out of the bloodstream and into tissue, probably in response to the body's efforts to eliminate the bacteria from the bloodstream. The bacteria may lie dormant for a period of months or years, but they are not vanquished. In some people, the bacteria trigger serious neuromuscular damage, and a painful death can ensue.

Some moralists living at the time when syphilis ran its course unimpeded would have disapproved of a cure for syphilis, on the basis that syphilis was one of the "wages of sin." These people seemed not to have realized, or did not care, that syphilis took a toll on innocent bystanders as well as sinners. For example, wives or husbands who led a perfectly virtuous, monogamous life could be infected by a straying spouse. Moreover, women infected with syphilis ran the risk of giving birth to malformed infants because the bacterium that causes syphilis is able to cross the placenta and infect the fetus.

Today, fortunately for both the innocent bystanders and sinners alike, syphilis is no longer the much-feared disease it once was. Antibiotics such as penicillin eliminate the bacteria in the early stages of the disease so that none of the adverse consequences of the disease develop. Consider, by contrast, the plight of a syphilitic patient in the bad old preantibiotic days.

In a book entitled *The Wages of Sin: Sex and Disease, Past and Present*, Peter Allen describes a widely used treatment of the 15th and 16th centuries. The patient was smeared with a cream containing a high concentration of mercury and shut into a small hut called a "stew," given that name because it was heated to induce sweating. Many patients died of this treatment, due to accidental overheating or the toxic effects of mercury. Those who survived lost teeth and developed sores in their mouths and throats. Despite their sufferings, few patients were cured of the disease. The fact that many syphilis sufferers would willingly seek out such a drastic and dangerous treatment gives a good indication of how desperate they were to be cured of the disease. More enlightened treatments, such as the ingestion of derivatives of arsenic, were introduced in the late 1800s, but this type of treatment also had unpleasant side effects and did not always cure the disease. Only with the widespread availability of penicillin in the 1940s did syphilis become an easily controllable disease. In fact, there are some public health officials today who dare to hope that syphilis might be eradicated entirely from the United States in the near future—thanks to penicillin.

In this book, many tales will be told of bacteria that have become resistant to antibiotics. In this respect, *T. pallidum* is something of an anomaly. It is one of the few disease-causing bacteria that has remained susceptible to penicillin, one of the earliest antibiotics, whereas most other disease-causing bacteria have become resistant to it. We will return later to the perplexing question of why some bacteria become resistant to antibiotics so much more readily than others.

Tuberculosis

Another dramatic example of the impact of antibiotics is provided by the lung disease tuberculosis. Tuberculosis is caused by a rod-shaped bacterium, *Mycobacterium tuberculosis* (Fig. 1.2). Although tuberculosis was much feared, it was considered by some to be a "romantic" disease

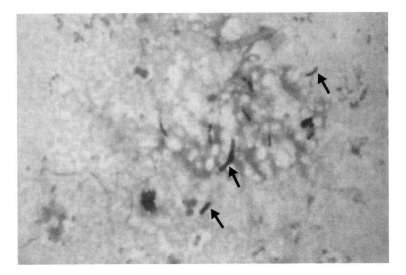

Figure 1.2   *Mycobacterium tuberculosis*, the cause of tuberculosis, as seen in a specimen from a patient. (Courtesy of George P. Kubica, CDC, Atlanta, Ga.)

because many artists and writers contracted it. Also, it caused an ethereal pallor in the sufferer, highlighted by a rouge-like coloration of the cheeks. Never mind that the breath of tuberculosis sufferers was foul enough to put off the most devoted admirer. At a distance a tuberculosis patient looked rather distinctive in an anemic sort of way.

What was happening inside the patient was the very opposite of romantic. Once inhaled into the lung, the bacteria begin to multiply. Many people can bring this initial infection under control, but in about 5% of those infected, the bacteria cause so much damage to the lungs that the patient coughs up blood and eventually dies. In some people, the bacteria move into the bloodstream and then into various organs, causing an even more aggressive and rapidly fatal disease once known as "galloping consumption."

There is another aspect of tuberculosis that deserves mention. In the case of syphilis, a person could avoid contracting the disease by living a monogamous lifestyle, unless of course one had the misfortune to be raped, to have an unfaithful spouse, or to be the child of a syphilitic parent. In the case of tuberculosis, by contrast, the disease is acquired by inhaling bacteria introduced into the air by people with active disease. In other words, the main risk factor for acquiring tuberculosis is breathing. This airborne mode of transmission makes it much more difficult for a person to protect himself or herself from acquiring the disease.

Today, tuberculosis patients take a course of antibiotics to cure their disease. This course of therapy lasts for at least 6 months, unlike most courses of antibiotics, which are taken for days or weeks, and the antituberculosis antibiotics can have unpleasant side effects such as nausea or liver damage. Nonetheless, given the alternative of an agonizing death as one's lungs are destroyed by the bacteria, antibiotic therapy for tuberculosis clearly represents a major advance in the control of this disease.

To understand the difference the antituberculosis antibiotics have wrought, consider how you would have fared as a tuberculosis sufferer in previous centuries. Tuberculosis patients who could afford the most advanced medical treatments of the time went to sanitaria in the mountains, where they spent hours in the open air and ate rich meals in an attempt to counter the wasting of the body that accompanies the advanced form of the disease. Pleasant as this treatment might sound to today's spa enthusiasts, it was sadly unsuccessful in most cases. Though people who went to sanitaria had the luxury of dying in scenic surroundings while partaking of luxurious cuisine, they died nonetheless.

Thomas Mann, the Nobel Prize winning German novelist, described a sanitarium for tuberculosis sufferers in his novel *The Magic Mountain*. On the surface, the sanitarium seems like a delightfully luxurious hotel, but this appealing surface conceals the grim reality of the disease. One of Mann's main characters, a young tuberculosis patient called Hans Castorp, arrives at the sanitarium. Castorp is taken to his new room by his cousin Joachim, who is also a patient.

At first, Hans is delighted with his room, but Joachim nonchalantly exposes its dark history. " 'An American woman died here day before yesterday,' said Joachim. 'Behrens [a sanitarium administrator] told me directly that she would be out before you came . . .Night before last, she had two first-class hemorrhages, and that was the finish. But she has been gone since yesterday morning, and after they took her away of course they fumigated the room thoroughly with formalin, which is the proper thing to use in such cases.' " Unfortunately, although formalin (formaldehyde, a general poison that kills bacteria as well as people) may have prevented acquisition of the bacteria from inanimate objects, it did nothing to prevent transmission through respiratory droplets from other patients.

Less wealthy people, who could not afford to go to luxurious mountain sanitaria, had their own low-cost version of the same therapy. Whole families lived in open wagons, seeking refuge in the fresh air "cure." Others simply perished miserably in their homes or rooming houses.

Perhaps the most inventive, if unsuccessful, strategy for treating tuberculosis was devised by rural farmers in the Northeastern United States in the 1700s, a strategy that reveals the desperation that people of the time felt when they or members of their families faced this dreaded disease. This particular treatment was based on the theory that tuberculosis was caused by vampires. This theory is not as laughable as it might seem at first. Victims of tuberculosis often had close relatives who had previously succumbed to the disease and who, after death, might have returned to prey on the living. Also, as the disease progressed, disease sufferers became deathly pale and ravenous.

The therapy that flowed logically from this understanding of tuberculosis was to disinter people who had recently died of the disease and rearrange their bones to make it difficult for them to rise from their graves and walk among the living. Separating the skull from the spinal column and crossing the leg bones across the sternum were common rearrangements. Apparently, vampires were forgetful creatures and did not remember enough anatomy to correct these rearrangements of their body parts.

This attempt to treat tuberculosis may seem amusing to us today, but in the absence of information about how tuberculosis is spread and effective therapy to treat it, this was the best people could do. Desperation provides a powerful stimulus to the imagination.

Today, tuberculosis is still a scourge worldwide and has even enjoyed a resurgence in some developed countries that have let down their guard. Nearly one-third of the people in the world are infected with *M. tuberculosis*, and millions of those infected people develop the full-blown disease and die each year. Tuberculosis has been a particularly tragic disease in places like Africa, with its high rate of human immunodeficiency virus (HIV) infection, which debilitates the immune systems of infected people and makes them more likely to die of tuberculosis. Some have called HIV-tuberculosis the "one-two" punch of death. So, unlike syphilis, tuberculosis is not likely to be eradicated any time soon, even in developed countries, and antituberculosis drugs accordingly remain critically important.

Bacterial pneumonia
Stories rooted in past centuries often have less of an impact on us modern folks than stories of diseases that are more immediate threats. An example that takes us into the 20th century was the first successful treatment of another lung disease called bacterial pneumonia. This type of lung disease is most commonly caused by a spherical bacterium, *Streptococcus pneumoniae* (Fig. 1.3).

Cells of *S. pneumoniae*, which are often seen in pairs (diplococci), may look like tiny kissing beach balls, but *S. pneumoniae* is trouble with a capital T. Bacterial pneumonia is currently the most common cause of infectious disease deaths in the United States, more common than HIV/AIDS, severe acute respiratory syndrome (SARS), or the other diseases du jour. Paradoxically, the high incidence of bacterial pneumonia may be part of what causes it to be overlooked by the media. Where is the news in a disease that routinely kills tens of thousands of people a year? The fact that many of these people are elderly adds to the story's lack of newsworthiness, except of course for those who are moving into their golden years.

Point out, however, that the second-most-common sufferers of pneumonia and meningitis caused by this bacterium are infants, and even the most hardened cynic might begin to take notice. *S. pneumoniae* is also the most common cause of earache in children. Although this form of *S.*

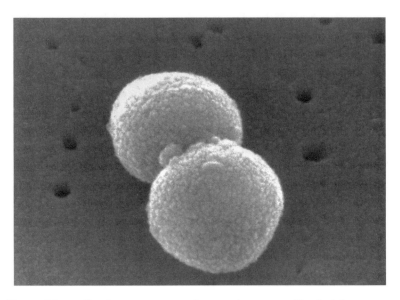

Figure 1.3   *Streptococcus pneumoniae,* a major cause of bacterial pneumonia, as seen with an electron microscope. (Courtesy of Janice Carr, CDC,
Atlanta, Ga.)

pneumoniae infection is seldom fatal, it disrupts the lives of many families,
especially those with children who undergo repeated episodes of earache.
The deadliest form of *S. pneumoniae* in infants is infant meningitis, which,
if it does not kill, may leave the survivor blind, deaf, or brain damaged.
Until recently, diseases caused by *S. pneumoniae* responded well to penicillin in most cases, but resistance to penicillin has been increasing and treatment failures in the case of earache cases are becoming more and more
common.

A story that illustrates the dramatic effect of the introduction of antibiotics was told recently in a *New York Times* obituary. In June 1999, Anne
Sheafe Miller died at age 90, after a long, rich life. An ordinary citizen,
her obituary was nonetheless accorded rare two-column coverage in the
New York Times. The reason? Anne Miller was the first American citizen
whose life was saved by penicillin (Fig. 1.4). According to the *Times* account, Mrs. Miller was lying near death in a Connecticut hospital, delirious
and spiking temperatures as high as 107°F. She had been hospitalized for
a month, treated with sulfa drugs (an early antibiotic and the best therapy

Figure 1.4   Anne Sheafe Miller with her doctor (left) and Sir Alexander Fleming, the discoverer of penicillin (right). (Courtesy of the family of Anne Sheafe Miller.)

of the time), transfusions, and surgery. Unfortunately, these draconian attempts to bring her infection under control were not working.

In a last-ditch attempt to save her life, doctors injected her with a new drug—penicillin—which was rushed to the hospital from a New Jersey pharmaceutical laboratory. Within 24 h, she was no longer delirious, her temperature had returned to normal, and she was eating normally. As the *Times* article put it, "news of Mrs. Miller's full and seemingly miraculous recovery helped inspire the American pharmaceutical industry to begin full-scale production of penicillin." The real news in the *Times* obituary, however, was that because Mrs. Miller survived certain death as a young adult, she was able to live to the advanced age of 90.

Surgery and cancer chemotherapy made safe
Another example of the power of antibiotics began with World War II. In the preantibiotic period, wounded soldiers who did not die immediately from the trauma caused by their wounds frequently developed deadly wound infections caused by such bacteria as *Staphylococcus aureus*

and *Streptococcus pyogenes* (the bacterium whose most recent claim to fame was its appearance as "the flesh-eating bacterium"). Amputation was the treatment used in cases of infections that occurred in the arms or legs, but even this drastic intervention had only limited success in saving the patient.

World War II was the first war in which wound infections were not the major cause of amputations and death. The reason was the newly available antibiotic penicillin. In fact, there is a famous poster from World War II that depicts a fallen soldier being held by a medic. The caption reads, "Thanks to PENICILLIN . . . he will come home" (Fig. 1.5).

Even in times of peace, wound infections have continued to be a serious medical problem in another context, surgery. Physicians have known and practiced surgical procedures for centuries, but only with the discovery of disinfectants and antibiotics did surgery become the relatively low-risk procedure it is today. Unfortunately, the bacteria that cause postsurgical infections have been becoming increasingly resistant to antibiotics, and some appear to be becoming resistant to disinfectants, a development that should be causing alarm in medical circles. Increasing exposure of patients to infection and increasing resistance of bacteria to antibiotics not only prolong the stay of surgical patients in the hospital, thus increasing medical costs, but could also make surgical procedures that are now considered routine too risky to be performed except in emergencies.

Still another area in which antibiotics have been important adjuncts of medical treatment is cancer chemotherapy. Currently available anticancer therapies resemble in some respects the old treatments for syphilis, in that they do not target the problem (cancer cells) specifically but attack all rapidly dividing cells of the body. Among these rapidly dividing cells are the neutrophils, white blood cells that are a major first-line defense against bacterial infections.

Patients who are receiving aggressive cancer chemotherapy experience a drastic decrease in these white cells during the course of their therapy. As already mentioned, to prevent such patients from succumbing to overwhelming bacterial infections, physicians routinely prescribe antibiotics to be taken during the period when the white cell count is lowest. This preventive therapy tides the patient over until the white cell counts return to normal.

The bottom line is that, in addition to curing such once-dreaded diseases as syphilis, tuberculosis, and bacterial pneumonia, antibiotics have

Figure 1.5 World War II poster extolling the virtues of penicillin. (Courtesy of Pfizer Inc.)

also become a major factor in making many modern medical procedures possible. Many of the modern medical miracles, ranging from surgery to cancer chemotherapy, rest on the continuing efficacy of antibiotics. Without antibiotics, these techniques would still be available, but they would be much riskier than they are at present.

More modest but important gains: ulcers and acne

Not everyone with an infectious disease dies. Many infectious diseases merely (merely?) diminish the quality of life. Two examples of this phenomenon are acne and ulcers. There is still controversy about the cause of acne, but there seems to be no question that antibiotics such as oral tetracycline and topical clindamycin can have a dramatic ameliorating effect on such skin diseases as acne and rosacea (a reddening of the skin, like a blush, that does not fade).

The ulcer story is more clear cut. For years, physicians were convinced that gastric ulcers were caused by stress. Supposedly, stress led to increased output of stomach acid, resulting in the formation of painful ulcers in the lining of the stomach or duodenum. During the 1980s, a heretical idea began to be more widely accepted: that most gastric ulcers are caused by a spiral-shaped bacterium called *Helicobacter pylori* (Fig. 1.6).

The heresy finally gained acceptance as fact when an antibiotic regimen was developed that eliminated *H. pylori* from the stomach and, in the process, eliminated the patient's ulcers. For patients who had become inured to a bland diet and years of administration of expensive medications, which cured an existing ulcer but did not prevent development of new ulcers, the week-long course of antibiotics that eliminated future ulcers was a revolution. Granted, it was a small revolution compared to the control of syphilis and tuberculosis, but for people with ulcers it was pretty important.

There are many other conditions, such as infected skin wounds that are painful but not deadly and certain types of debilitating traveler's diarrhea whose duration can be reduced with antibiotics, that patients would rather have cured than endured. Another category of diseases that will be described in chapter 9 is diseases once thought not to be caused by bacteria but now suspected of having an infectious cause. Examples are certain forms of heart disease, preterm birth, and inflammatory bowel disease. This work is controversial but potentially exciting.

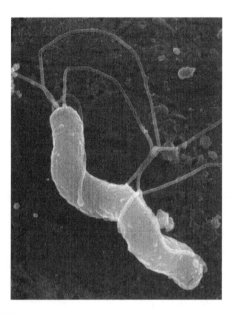

Figure 1.6   *Helicobacter pylori*, the cause of most gastric ulcers, as seen with an electron microscope. (Reprinted from J. O'Rourke and G. Bode, p. 53–67, *in* H. L. T. Mobley, G. L. Mendz, and S. L. Hazell, ed., Helicobacter pylori: *Physiology and Genetics*, 2001.)

Issues to Ponder

These questions, which will end most chapters, are questions for which there are no easy answers but that are meant to illustrate the continuing debate over how concerned we should be about resistance to antibiotics and what we should do about the problem.

1. There is a school of historians that has claimed that antibiotics and vaccines have had at most a minimal effect on human health. They cite numbers showing that the incidence of infectious disease deaths was declining significantly well before the introduction of antibiotics and vaccines.

No one disputes the fact that changing living conditions, such as reduced crowding (possibly augmented by immunity developed as a result of survival of devastating plagues), had a major effect on increased human survival as early as the 16th century. In this chapter, we present a role for antibiotics that is more dramatic than this view would admit.

Is there a fundamental contradiction here, or is this really a difference in perspective on human health between a view that focuses on the population as a whole and a view that stresses the individual?

2. What is the real importance of antibiotics, if any, to you? To the population at large? Keep in mind that your chances of contracting a serious infection are relatively low, certainly lower than your chance of dying in an automobile accident. How important is it to you personally to act in ways that help preserve the efficacy of antibiotics?

3. From time to time, the media have raised the specter of a return to a preantibiotic era. Suppose the worst happened and all antibiotics were to become ineffective. Would health care return to the level experienced by people living before the 20th century or are there some ameliorating factors based on what has been learned in the past century, such as improved hygiene and disinfectant use? What changes in medical culture might have to be made to maximize the effects of such ameliorating factors? This topic is discussed in chapter 4.

2

A Brief Look at the History of Antibiotics

Streptomycin, sulfa drugs, and penicillin were the first antibiotics to move from the laboratory into general clinical use. Subsequently, scientists have developed many new forms of these antibiotics and are currently looking for new types of antibiotics with targets different from those of the older antibiotics, but as will become evident, this quest is not as easy as it was in the early years of antibiotic discovery and development.

Purifying Soil and Beautiful Earwax

Antimicrobials have not always been as user-friendly as they are now. Today's antibiotics have undergone extensive testing and approval procedures. As mentioned in the first chapter, the earliest antibacterial compounds, such as mercury and the derivatives of arsenic, were almost as toxic for us as for the bacteria. This early approach to therapy was based on what has been called the poison principle. That is, known poisons were administered in limited doses in the hope that the bacteria causing the infection would be killed before the person being treated.

Rene Dubos, a research scientist at Rockefeller University, was the first to take a very different view of how antibacterial compounds should work. His view emerged naturally from his lifelong study of soil microbes. Dubos had an almost religious reverence for the purifying properties of soil. In an experiment that was destined to make history, he targeted the bacterium *Streptococcus pneumoniae*. *S. pneumoniae* is the most common cause of bacterial pneumonia and has been a major killer for centuries. Dubos' strategy was to mix a laboratory culture of *S. pneumoniae* with an aqueous extract from soil. His theory was that there must be bacteria in soil that could kill or inhibit the growth of *S. pneumoniae* because the ecological balance would be maintained only if such microorganisms existed. His idea made sense because *S. pneumoniae*, despite its ability to colonize the human body and cause serious disease, has not taken over the world and is very uncommon in soil.

Dubos isolated a soil bacterium, *Bacillus brevis*, which produced a substance that was antagonistic to the growth of *S. pneumoniae*. Dubos may not have realized it at the time, but he was on the verge of a new paradigm for fighting bacterial infections and a new era of medicine.

Unfortunately, in the beginning of his quest, all was not sweetness and light. The initial form of his antibacterial substance had some rather unappealing characteristics. As Rollin Hotchkiss, a colleague of Dubos, later described it in the book *Launching the Antibiotic Era* (Rockefeller University Press, 1990) the antibacterial compound was a "crude brownish material [that] . . . congealed into a sticky mass as unpleasant as so much uncouth earwax. But it was a powerful wax all right." An extract from this "uncouth earwax" was able to inhibit the growth of such bacteria as *S. pneumoniae*. Ultimately, Hotchkiss and others isolated the active component of the uncouth earwax, a compound we now know as gramicidin.

Gramicidin is a peptide that forms channels in bacterial membranes. Because the cytoplasmic membranes of bacteria and humans are very similar in composition, gramicidin proved to be too toxic for internal use in humans, although not so toxic as mercury and arsenic. It is still used as an ingredient in topical antibacterial preparations. The importance of the discovery of gramicidin was that it directed attention to soil microbes as a possible source of antibacterial compounds. Subsequently, soil bacteria and fungi proved to be rich sources of antibiotics, such as penicillin and tetracycline, that were much more human friendly than gramicidin.

The Sulfa Drugs

Parallel to the quest by Dubos and his colleagues for natural products of soil microbes, another line of research was emerging. This approach was to modify compounds that kill bacteria to make these compounds less toxic for humans. This approach had been tried with arsenic, but the derivatives of arsenic were still too toxic and were not very effective against bacteria. As chemists became more sophisticated, however, they experienced their first success: synthetic compounds called sulfonamides.

The discovery of sulfonamides arose from the observation that a red dye, Prontosil, could cure some cases of pneumonia. In the early 1930s, scientists discovered that the active component in Prontosil was a compound that was converted by human cells into an antibacterial compound called sulfanilamide. Sulfanilamide was not nearly as toxic to humans as mercury and arsenic. Thus were born the drugs that came to be known as "sulfa drugs."

We now know that sulfanilamide and other sulfa drugs mimic para-aminobenzoic acid, a precursor of the vitamin folic acid. Bacteria make their own folic acid, whereas humans obtain preformed folic acid from their diet and thus do not have to synthesize it themselves. Because of this difference in metabolism, chemical mimics of para-aminobenzoic acid affected bacteria adversely by inactivating an enzyme the bacteria needed to make folic acid, but they had no effect on human cells, which do not have to produce folic acid from such precursors. From these two parallel lines of research emerged a new principle: the principle of selective toxicity, toxicity against bacteria but not against humans.

Penicillin Is Discovered (Almost by Accident)

There has been a longstanding debate about who actually discovered penicillin, the antibiotic that, despite the early successes of the sulfa drugs, unquestionably gave the antibiotic revolution its huge momentum. Some historians credit Alexander Fleming, at Oxford University in the United Kingdom, as the discoverer of penicillin. His contribution arose from a series of experiments with a bacterium that was a notorious cause of life-threatening wound infections, *Staphylococcus aureus* (Fig. 2.1). Fleming

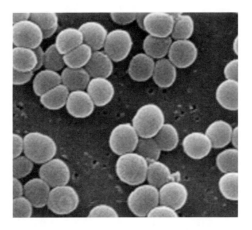

Figure 2.1 *Staphylococcus aureus*, a cause of serious wound infections, as seen with an electron microscope. (Courtesy of Janice Carr, CDC, Atlanta, Ga.)

noticed that on some agar plates that had been inoculated with *S. aureus,* which normally forms colonies over most of the surface of the plate, there was an inhibitory zone in which no bacteria grew. This zone had developed around a colony of what turned out to be a fungus, *Penicillium notatum,* which was later identified as the producer of penicillin. However, this discovery was not as intentional as textbooks tend to imply.

Alexander Fleming was a microbiologist-physician who was interested in a more effective treatment for wound infections. Fleming had focused on *S. aureus* because of its role in war wound infections. He had some interest in new compounds that might be used to control bacterial infections, but his primary focus had been on known antibacterial compounds such as arsphenamine, a derivative of arsenic that had attracted much attention because of its success in curing some cases of syphilis. However, arsphenamine was a rather toxic compound.

Fleming knew that some of the bacteria with which he was working could be dangerous. Accordingly, he discarded used agar plates containing colonies of *S. aureus* into trays filled with a disinfectant that was supposed to kill the bacteria. Like some microbiologists of the time, however, he could be careless with his discarded specimens.

Read an account of what actually happened, written by Norman Heatley, one of the early workers in the area of antibiotic research (*Launching the Antibiotic Era*).

> In the summer of 1928, Fleming goes on holiday, unaware that he has been chosen by the Fates to take the first steps in introducing the antibiotics to mankind. Having made a wise choice of their agent, the Fates also arranged that one of his plates, inoculated with staphylococci but not incubated [at 37°C], should be contaminated with a spore of [the fungus] *Penicillium notatum,* and that the weather conditions during the subsequent weeks should provide the sequence of rather narrow temperature ranges required to produce the penicillin effect [killing of surrounding bacteria]. Fleming returns from his holiday and goes through the pile of used plates on his bench, looking at them and discarding them into the tray of disinfectant. The plates are numerous and soon they pile up, above the disinfectant. But what is this? Gracious heavens, he has discarded *the* plate! All is not lost, for the Fates have a messenger on hand in the form of Fleming's colleague, D. M. Pryce. Pryce makes his entrance, they chat about staphylococci and to make a point, Fleming picks up some of his discarded plates. The Fates hold their breath. Yes! He picks up *the* plate, looks at it, and says "That's funny . . ." [See what Fleming saw in Fig. 2.2]. How fortunate that trays rather than buckets were used for discarded cultures and that D. M. Pryce was on hand at the critical moment.

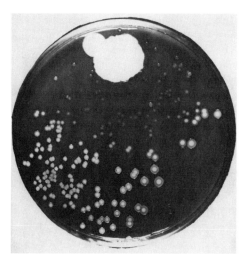

Figure 2.2   Fleming's plate of *Staphylococcus aureus* colonies, which shows an area of clearing around the much larger colony of *Penicillium notatum* at the top of the plate. © Bettmann/CORBIS.

Although Heatley and Fleming clearly viewed Fleming's careless laboratory practices as a gift from the gods, microbiologists today would see that carelessness differently. The events described in Heatley's account, particularly the stacks of plates rising out of the disinfectant, would have gotten Fleming in a lot of trouble if he were practicing microbiology today, possibly leading safety officers to shut down his laboratory until he cleaned up his act. Laboratory safety is taken a lot more seriously today than it was in Fleming's time. In fact, a book on legally mandated safety procedures for microbiological laboratories, which is published by the Centers for Disease Control and Prevention, now exceeds 200 pages.

Does this mean that bureaucratic interference is currently stifling great scientific discoveries? Obviously not, since the pace of progress in microbiology has increased a lot since Fleming's day. It just means that scientists now have a better appreciation for the safety of fellow laboratory workers and the public than they did in the early days of microbiology. Also, good laboratory procedures make for more reliable, more believable scientific results. The experience of microbiologists in the ensuing decades has convinced us that the kind of sloppiness exhibited by Fleming is far more likely to lead to bad science than to great discoveries.

Although Fleming is often credited with the discovery of penicillin, many would argue that the scientists who deserve credit for the real successes of penicillin were an Australian scientist, Howard Florey, and his assistant, Ernst Chain. Florey and Chain were scientists who did not see penicillin as merely another microbiological curiosity, as Fleming did initially, but instead dedicated themselves to discovering how to produce enough penicillin to make the drug widely available. Prior to their discoveries, penicillin was available only in very limited quantities. Without their intervention, penicillin would have had only modest impact. Only when it was produced in large scale did penicillin begin to lead to mass cures of diseases from wound infections to syphilis to bacterial pneumonia.

It is comforting to see that in some ways virtue is rewarded. Even if Florey has not gotten the credit some people think he deserves, his fellow Australians have recognized his contribution in a very material way. His picture appears on the Australian 50-pound note. Even the Queen of England merits only the 5-pound note.

In the race to produce enough penicillin to meet military and civilian demand, scientists resorted to a variety of growth vessels, ranging from glass bottles that had originally contained popular drinks to bedpans. The goal was to grow large amounts of the fungus *P. notatum* in order to harvest the culture liquid that contained the antibiotic secreted by the fungus. The main problem was not just volume, however, but that culture supernatants from *Penicillium* had very low activity.

Later, Chain realized that the low potency of the culture fluid was due to the fact that the initial cultures of *Penicillium* were contaminated with a strain of the common bacterium, *Escherichia coli*. This particular *E. coli* strain produced an enzyme that degraded penicillin. Thus, even at the earliest steps in the discovery of what would ultimately become one of the most important antibiotic classes ever discovered, the penicillin family of antibiotics, scientists also saw the first evidence that bacteria could become resistant to penicillin. Uh oh.

Contaminating the Miracle: The Tuskegee Study

Discovering antibiotics and producing them in quantities sufficient to be useful to the population at large were only a part of the process. Making sure that these products were equitably distributed was equally critical. Anyone who has been present at graduation ceremonies for medical students about to enter their profession can hardly fail to be moved by the Hippocratic oath, an important part of which is "above all, do no harm."

What happens when physicians forget this oath? Unfortunately, we have an answer to this question in the form of the so-called Tuskegee study.

Syphilis was widespread in the southern United States in the 1940s, as it was in many other parts of the world. The discovery that arsenicals could successfully treat some cases of syphilis pointed to a breakthrough in the treatment of this horrible disease. The success of the arsenicals was spotty, however, so it is perhaps understandable that although those rich enough to afford this new treatment were able to get it, there was not a massive public health assault on the disease in communities where people were too poor even to afford visits to doctors. Then penicillin came along, offering a gentle cure that was highly effective, and it was soon cheap enough to allow a massive campaign to eradicate syphilis, even among impoverished populations.

To make a long sordid story short, physicians associated with the U.S. Public Health Service, the precursor of the current U.S. National Institutes of Health, decided that there was not enough information on the pathology of syphilis to complete the textbook description of the disease. Their solution to this problem was to create a study in which a group of syphilitic African American sharecroppers in Alabama would not be given the new antibiotic treatment. Instead, they would be offered free "health care" that consisted mostly of collecting blood samples and cataloging the deterioration of their health. Since these men had never had any health care at all, they were easily convinced that they were lucky to be included in a study where at least they saw a nurse on a regular basis. It was only during the 1970s that this shameful study was exposed and the few men in the study who were still alive were finally treated and compensated for their suffering. Many classic political cartoons appeared when this outrage was revealed. A number of them are included in the book *Bad Blood* by J. Jones (Simon and Schuster, 1993).

To withhold effective antimicrobial therapy from sick people would be unthinkable today, right? Well, perhaps not. Most of the people suffering from tuberculosis in the world today will not receive the (very cheap) antibiotics that could cure them because they cannot afford those antibiotics and no one with money cares enough to provide them.

Perhaps the highest profile recent controversy in this area, however, has been the one surrounding azidothymidine (AZT). AZT is an anti-human immunodeficiency virus (HIV) compound that does not cure HIV infection but at least slows the progression of the infection to AIDS and has prolonged the survival of many AIDS patients. In particular, short-term AZT administration to pregnant women about to give birth is now known to prevent the transmission of HIV from the infected mother to the infant. We know this because of the results of a very controversial set of clinical trials.

The short-term AZT therapy was tested in Africa and Thailand, but any ethics panel in a developed country would not have approved the design of this clinical trial. Two populations were compared. One group of women was given the short-term course of AZT, and a second group, the control group, was given a placebo. In the United States and Europe, the comparison would have been between women given the new short-term AZT regimen and women given the much longer-term regimen now used in developed countries to prevent mother-to-infant transmission.

The AZT trials in Africa and Thailand caused a storm of international protest and raised the inevitable suggestion of parallels to the Tuskegee study. Unquestionably, the two studies differ vastly from each other. The AZT trials were designed and carried out by physicians and scientists who were dedicated to helping prevent HIV transmission and who had the health of the subjects in the study very much in mind. However, the AZT trial controversy shows that the ethics of drug testing can still be a contentious issue, especially when trials are conducted in poor countries.

It is now very expensive to test new antibiotics for safety and efficacy in countries like the United States. It is hard to imagine that any reputable pharmaceutical company would try to cut costs in the way they were curtailed in the case of the short-term AZT trials, but only constant vigilance by the scientific community will ensure that the testing and distribution of precious, life-saving antimicrobial drugs will be equitable and ethical.

Issues to Ponder

1. Who really does deserve the credit for the discovery of penicillin: Fleming or Florey and Chain? This is actually a fairly deep question about assigning credit for a scientific discovery. The credit for such a discovery usually goes to the person or persons who first noticed a phenomenon and performed the first experiments to test the hypothesis the phenomenon suggested. However, effective delivery of such insights to the public is also an issue for all of us. In the case of a work of art, the curators and exhibitors do not get the same degree of credit as the artist who created the work, but in the case of a medical treatment like penicillin, how useful

is a drug that remains a curiosity and never makes it to the people who need it?

2. As will be described in chapter 9, the declining profitability of antibiotics has made them less and less desirable to pharmaceutical companies. Also, antibiotics are old news to the scientific community. Accordingly, research on such promising but so far nonproductive therapies as gene therapy and stem cell research have taken the publicity foreground in recent years. No one disputes the importance of giving high billing to promising areas of research that need all the encouragement they can get. However, the relative lack of interest in antibiotic discovery on the part of scientists and the public may not be such a good idea if the goal is to encourage development of new antibiotics to meet the challenge of increasingly resistant bacteria. What should be the relative interest in and coverage of exciting new, but still unproven, therapies compared to older successful but no longer as exciting therapies? This is not a simple question, because it asks about the balance between encouraging innovation and supporting the continuation of past successes.

3. Today, there is no question that the Tuskegee study was a moral abomination. Nonetheless, it is still troubling that it happened at all, and, even more disturbing, that the controversy raised by Tuskegee was echoed, however faintly, in the recent AZT trials. In medical school programs, lectures on ethics are now de rigueur, but is the Tuskegee case part of that training? In most medical schools, the answer has been "no," and it is uncommon to find medical students today who even know this story. There are two schools of thought on the issue of whether to include this historical account in a microbiological curriculum. One, which we advocate and thus will present first, is that the Tuskegee experience is such a stark example of how highly trained scientists can go astray that it grabs the attention of students, making them, one hopes, more sensitive to ethics discussions that have more subtle shadings. The countervailing view is that the Tuskegee experience was so starkly an example of criminal behavior that it will never be repeated and thus does not merit discussion today. What do you think? Remember, medical students are being bombarded with an expanding mass of new medical information, which makes time available for coverage of any subject more and more precious.

3

Bacteria Reveal Their Adaptability, Threatening the Brief Reign of Antibiotics

The great medical progress made possible by antibiotics, progress that was celebrated in chapters 1 and 2, has not been without its problems. In recent years, bacteria have fought back by taking advantage of their own adaptive powers to become resistant to antibiotics.

Why the Appearance of Bacteria Resistant to Antibiotics Was a Forgone Conclusion

The rapidity with which some bacteria have developed mechanisms for resisting antibiotics should not have come as a surprise. Bacteria were one of the first forms of life on Earth. They ruled the Earth for over 3 billion years before insects, plants, and animals began to appear. During that period, they colonized every part of the Earth, from the deepest part of the ocean to the highest mountains. They can be found under ice in the Arctic and Antarctic regions. They can be found in boiling hot springs, in nuclear wastes, and as far underground in the land masses as humans have been able to dig. They have even been found in clouds and inside rocks. They survived the volcanic era of Earth's early days, as well as periodic fluctuations in temperatures ranging from ice ages to greenhouse conditions. Is it any wonder, then, that antibiotics represented at best a minor nuisance in the long march of their evolution, a nuisance that could easily be brushed aside?

For us humans, the ability of bacteria to become resistant to antibiotics poses a serious problem. We can, and will, try to keep pace with the development of bacterial resistance by developing new antibiotics, but is it possible that we are in a race with bacteria that we will ultimately lose? Could we one day return to a preantibiotic era? This frightening possibility led an anonymous author to write the following terse account of one view of our past, present, and future success in combating infectious diseases:

- 2000 B.C.—Here, eat this root.
- 1000 A.D.—That root is heathen. Here, say this prayer.
- 1850 A.D.—That prayer is superstition. Here, drink this potion.
- 1920 A.D.—That potion is snake oil. Here, swallow this pill.
- 1945 A.D.—That pill is ineffective. Here, take this penicillin.
- 1955 A.D.—Oops . . . bugs mutated. Here, take this tetracycline.
- 1960–1999 A.D.—39 more "oops" . . . Here, take this more powerful antibiotic.
- 2000 A.D.—The bugs have won! Here, eat this root.

In this book, we will strive to present a more optimistic view of our future, but public concern about the declining efficacy of antibiotics and where this trend might lead is understandable. Even the more optimistic among us still believe that the battle between humans and bacteria is far from over. The best we can hope for is detente, a running standoff. But that would be far better than losing antibiotics altogether. The keys to future success in saving antibiotics are knowledge and the willingness of the public to take an informed interest not only in preserving the efficacy of the antibiotics we have now but also in ensuring that future development of new antibiotics continues.

A caveat is in order here. Although we have just succumbed, as many people now do, to describing the continuing interaction between humans and bacteria as a war, it is important to point out that this view may well be a dangerous misconception. Only a small number of bacteria cause infections. Most bacteria are either neutral or benign from the human point of view.

The idea that the best strategy for maintaining human health is to kill all bacteria is not only stupid but suicidal, because without bacteria there would be no human life on Earth. Bacteria are important as recyclers and as the base of the Earth's food chain. Fortunately, even with the great ingenuity we humans possess, it is highly unlikely that our efforts will have any appreciable effect on the microbial world as a whole. Nonetheless, people continue to try to eliminate bacteria from their environment with antimicrobial soaps, wipes, and sprays, not realizing that these measures may be part of the disease problem not part of the solution.

The winning strategy, according to us, is to know your adversary.

Knowing your adversary means focusing on the minority of bacteria that cause disease and striving to prevent their incursions.

How Do Bacteria Become Resistant to Antibiotics?

The strategies bacteria use to resist the effects of antibiotics will be dealt with in more detail in chapters 5–7, which describe individual antibiotics, but it is worth taking a moment at this point to consider in general some ways in which bacteria might counter the action of an antibiotic that was previously capable of stopping their growth. Bacteria have evolved three types of strategies for circumventing the action of antibiotics.

First, the bacteria can destroy the antibiotic before it hits their vulnerable parts, a kind of antiballistic missile approach. Bacteria do this far more effectively and cheaply than humans, by producing specific proteins that chemically modify the antibiotic to a form that no longer interferes with the bacterial activity the antibiotic was designed to inhibit.

A second bacterial strategy is to take advantage of the fact that any compound that binds to a bacterial target, with a view to stopping its action, has to reach a threshold concentration in order to bind effectively. For example, a few stray antibiotic molecules are not going to be enough to stop a bacterial ribosome from synthesizing bacterial proteins. Thus, if a bacterium can keep the antibiotic concentration low enough, the antibiotic will not be able to do its job.

One form of this strategy is for a bacterium to insert a protein pump into its cytoplasmic membrane, the membrane that keeps the bacterium's insides in and the external environment out. This strategy might be called the revolving door strategy. An antibiotic that inactivates bacterial ribosomes, for example, must pass through the cytoplasmic membrane and enter the bacterial cell, rising in concentration to the point where antibiotic molecules bind effectively to most of the ribosomes in the cell.

If, however, the bacterial protein pump can eject the antibiotic from the cytoplasm as rapidly as the antibiotic moves into the cytoplasm, the concentration of antibiotic in the vicinity of the ribosomes will be too low to be effective in stopping the synthesis of bacterial proteins. Such protein pumps are called efflux pumps. Interestingly, although this mechanism of drug resistance was first discovered in antibiotic-resistant bacteria, it has now been found in human tumor cells, which use the same strategy to become resistant to chemotherapeutic agents.

A third bacterial strategy is to chemically modify or mutate the target of the antibiotic so that the antibiotic no longer binds. For example, some bacteria become resistant to penicillin by mutating the enzymes that penicillin inhibits, enzymes that are essential for forming the rigid cell wall of the bacteria. Other bacteria prevent antibiotics from harming their ribosomes, which are essential for synthesizing bacterial proteins, by modifying some component of the ribosome so that the antibiotic can no longer bind to the ribosome. This strategy can be somewhat dangerous for bacteria. Antibiotics are designed to interfere with bacterial components and processes that are essential for bacterial survival. Mutating or otherwise modifying such a component or process can have deadly effects on the bacterium. Imagine mutating your heart at random with a view to protecting yourself against heart attacks. Some mutations might succeed, but most of them would be lethal because they would disrupt the function of the heart. Yet bacteria have succeeded in finding mutations that allow them to resist the action of the antibiotic while still carrying out their normal life processes.

Is It Always the Bacterium's Fault? Other Reasons for Treatment Failures

Although the emergence of bacteria that have developed new ways to resist antibiotics is unquestionably a serious clinical problem, the failure of an antibiotic to cure your condition may not always be due to bacterial resistance. Physicians need to keep this possibility in mind, because moving to the newest high-powered (read "expensive") antibiotic may not be the best solution to a treatment problem. At the risk of being labeled bleeding-heart bacterium-protecting liberals, we venture fearlessly into this aspect of antibiotic treatment failures.

How can an antibiotic treatment fail for reasons other than the presence of bacteria that resist the action of the antibiotic? A major reason for apparent antibiotic failure is misdiagnosis of the infection. If the patient's condition is caused by a virus or a fungus, for example, and not by a bacterium, antibiotics will have no effect. An example of this can be seen in the case of some urinary tract infections. The vast majority of urinary tract infections are caused by bacteria. Thus, antibiotics are a standard treatment for the painful urination and fever associated with such an infection. In some people, however, urinary tract infections are caused by the yeast *Candida albicans*. Such infections are uncommon and occur mainly in people who have indwelling urinary catheters to drain their bladders or people who are immunocompromised. Such an infection cannot be treated successfully by an antibiotic, however. Similarly, misdiagnosis of a lung or genital tract infection, which usually is caused by bacteria but in some cases may be caused by fungi or protozoa, can lead to treatment failure if an antibiotic is prescribed as the treatment. Many scientists are coming to believe that development of new rapid diagnostic techniques could help to make sure that antibiotics are used only when they are likely to be effective.

Use of an antibiotic that has the wrong pharmacokinetic properties can also be a problem. "The wrong pharmacokinetic properties" is a fancy way of saying that the drug does not get where it is needed. This is not a common problem, because most physicians are well trained in the pharmacokinetics of the drugs they use, but it can happen. For example, an antibiotic that does not penetrate the blood-brain barrier (the membranes that cover the brain and spinal cord and separate spinal fluid from blood) will not do a meningitis patient, who has bacteria in the cerebrospinal fluid, much good.

Sometimes, the problem arises from incomplete information. For a long time, little was known about what types of antibiotics were most useful in treating bacteria in abscesses, where dead tissue inhibits the penetration of some antibiotics. Now, there is a better understanding of what antibiotics work best or whether removing the dead tissue is going to be essential before any antibiotic can enter the site.

Patients can play a role in treatment failure. Failure to take the full course of an antibiotic can leave viable bacteria to make a comeback when consumption of the antibiotic ceases. This problem has been most dramatically illustrated in the case of tuberculosis. The drug regimen for tuberculosis patients involves taking several pills. Many people have difficulty complying with this regimen. Additionally, many people have side effects such as nausea that they soon associate with some of the pills they are taking. Not only does intermittent taking of pills, or cessation of therapy entirely, leave live bacteria in the patient's lungs, but it increases the likelihood that the remaining bacteria may become resistant, rendering a renewed attempt at therapy ineffective.

Emerging Bacterial Resistance to Antibiotics Spawns New Generations of Antibiotics

Since the earliest antibiotics were discovered and introduced into clinical use, the pharmaceutical industry has continued to turn out new versions of old favorites as well as new types of antibiotics. In some cases, these new versions are designed to have improved pharmaceutical properties, such as better distribution in the body and fewer side effects. In most cases, however, the goal has been to modify the antibiotics to counter emerging bacterial resistance strategies.

New versions of antibiotics are described as if they were generations of a family. Thus, penicillin is called a first-generation antibiotic, whereas later alterations of the basic penicillin structure have been called second generation, third generation, and so forth. The multiplication of drug generations is not good news for people who will contract serious infections, because it bears testimony to the continued ability of bacteria to become resistant to each new generation of antibiotic. Also, each new generation bears a higher price tag because the newer antibiotics are still under patent protection.

The fact that we are currently in the fourth generation of most antibiotics, after only 60 years of antibiotic use, should be warning enough that the race between us and microbes is a hotly contested one. In chapter 9 we will explain why the major champions of the human side of the human-microbe race, the pharmaceutical companies, are finally tiring of the race for economic reasons and are dropping out. Will something replace these companies that pioneered the antibiotic revolution, will they be forced back into the race, or will we all give up and go back to eating the therapeutic, but ineffective, root? The answer is still not clear, but most of us would agree that somehow the antibiotic miracle needs to be sustained.

Until recently, the emphasis on maintaining antibiotic efficacy has focused on creating yet newer generations of antibiotics. However, scientists and physicians are now beginning to realize that another approach is needed—finding ways to prevent the development of bacterial resistance. Options include more prudent use of antibiotics to reduce the selection pressures that encourage resistant bacteria to emerge. Another strategy is directly targeting and inactivating bacterial mechanisms for resisting antibiotics.

An example of a drug used to counter a bacterial resistance mechanism is Augmentin, a popular drug for treating many kinds of infections. It will be described in more detail later on. The antibiotic component of Augmentin, amoxicillin (a member of the penicillin family), was falling out of favor because bacteria were becoming increasingly resistant to it.

Resistance was caused by bacteria beginning to produce an enzyme that inactivated amoxicillin. To counter this, Augmentin also contains a compound that inhibits the bacterial enzyme so that it can no longer destroy amoxicillin, leaving the amoxicillin to carry out its lethal task.

The Origins of Bacterial Resistance Genes: a Puzzling but Relevant Mystery

How do bacteria develop the special genes that make them resistant to antibiotics? Scientists are trying to answer this question by determining the DNA sequences of the genes that confer resistance and asking whether they resemble any other genes that were present before the resistance problem emerged. If there are similarities, perhaps scientists can construct a "fossil record" of the evolution of resistance genes that reveals how they arose. Also, did antibiotic-resistant bacteria emerge only after antibiotics were introduced into widespread clinical use or were they present before antibiotics were widely used?

So far, these investigations point to some tentative conclusions. First, a gene that makes a bacterium resistant to an antibiotic usually resembles a bacterial housekeeping gene, a gene that is responsible for an essential synthetic function such as building the rigid bacterial cell wall or contributing to the synthesis of bacterial proteins. These housekeeping genes appear to have mutated into genes that protect the bacterium from antibiotics that target the housekeeping gene.

A surprising finding has been the discovery that at least some antibiotic resistance genes were present long before humans began to produce and use antibiotics. It is important to understand this apparent anomaly because most strategies to slow the development of bacterial resistance to antibiotics rest on the premise that the current use of antibiotics is the only selective pressure driving the evolution and dominance of resistant strains. We hasten to point out that it is clear from many scientific studies that the current overuse of antibiotics is contributing in a significant way to the increased incidence of antibiotic resistant strains, but the nagging question persists: what if there are other selection pressures that we do not currently understand?

For a long time, the prevailing explanation for the existence of bacterial resistance genes before antibiotic use became widespread was that antibiotics were the first forms of germ warfare. That is, bacteria and fungi that produce antibiotics could use them to clear their area of competing bacteria, thus allowing the antibiotic producers to enjoy unmolested whatever nutrients or other benefits might be available in that particular location. There are two problems with this undeniably attractive idea.

First, no one has managed to detect antibiotics in soils where microbes that produce antibiotics in the laboratory are normally found. This finding is consistent with the fact that it is usually necessary to subject a possible antibiotic-producing microbe to mutagenesis in order to increase production of an antibiotic to industrially useable levels. As mentioned in the previous chapter, contamination of the antibiotic producer by other organisms that degrade the antibiotic can also be a problem, but enhancing antibiotic production by modifying the growth conditions and mutagenesis of antibiotic producers to increase production has also played a major role in maximizing antibiotic output.

Second, a basic principle of soil microbiology is that nature abhors a pure culture. If antibiotics were so effective in nature, there should be areas where a single species has succeeded in eliminating all competitors from its immediate vicinity. In general, the opposite is true.

A modification of the germ warfare hypothesis was suggested by the observation that the bacteria that produce antibiotics protect themselves by also having genes that make them resistant to the antibiotics they produce. Antibiotic-producing fungi have no need for this protection since they are eukaryotes, which are naturally not susceptible to antibiotic action. The interesting discovery that antibiotic-producing bacteria have resistance genes suggests a possible origin for antibiotic resistance genes but does not explain the fact that some bacteria that do not produce antibiotics have resistance genes too and apparently have had them since before humans developed antibiotics.

An interesting idea put forward recently by microbiologist Julian Davies at the University of British Columbia is that antibiotics are actually signaling molecules. In this view, resistance proteins would become receptors for the antibiotic signal or serve as proteins that enhance its activity. Fortunately, or unfortunately, depending on your bias, there is as little evidence to support this interesting idea as there is to support the germ warfare hypothesis.

Why is this debate of more than academic interest? Currently, scientists are designing strategies for controlling and slowing the increase in bacterial resistance to antibiotics on the assumption that the only selection for resistant bacteria is the use of antibiotics in modern medicine or in modern agriculture.

What if there are selective pressures other than antibiotic use by humans that are driving the evolution of antibiotic-resistant strains of bacteria? Ancient selective pressures might be enhanced by increased human influences on the environment such as pollution. There is some evidence that heavy metals and other pollutants may be selecting for bacteria resistant to antibiotics, a topic to which we will return in chapter 10. Such selective pressures might not be major contributors to the current rise in incidence of resistant strains but could be generating a constant source of small numbers of resistant bacteria, whose numbers increase dramatically if an antibiotic selection is applied. An understanding of such nonantibiotic selective pressures, if they exist, could have great practical importance in the battle to save antibiotics.

Issues to Ponder

1. We argue that antibacterial strategies should not aim to eradicate all bacteria but should focus instead on those causing the disease problem. Few physicians would disagree with this principle. Yet, in reality, when a physician prescribes an antibiotic to treat a specific infection, the antibiotic regimen affects all of the bacteria in and on your body. Your body is host to vast and complex bacterial populations on your skin, in your mouth, in your lower intestinal tract, and (if you are a woman) in your vagina.

Many people who take antibiotics experience the consequences of the effects of antibiotics on these natural populations. Two common consequences are diarrhea and yeast infections. In both cases a disruption of the normal population balance presumably allows disease-causing bacteria or yeasts to take over. Also, studies are beginning to show that treatment of a specific infection can increase the incidence of resistant bacteria in these natural populations. Given that members of these natural populations can sometimes give rise to postsurgical infections (a low probability event for most people), how seriously should this side effect be taken? Don't physicians already have enough to deal with without this type of complication?

2. This chapter lays out the usual bacterial resistance mechanisms, the ones for which there is abundant evidence: inactivation of the antibiotic, reducing the antibiotic concentration by pumping it out of the cell, and modifying the target of the antibiotic. Given that most of the antibiotic

targets are enzymes or other proteins that perform essential bacterial functions and given that many antibiotics act on targets in the cell cytoplasm, can you think of other ways bacteria might become resistant to antibiotics? Just because other mechanisms have not been described yet does not mean they do not exist. There is still a lot we do not know about the seemingly endless ability of bacteria to evolve around antibiotics.

4

Antibiotic-Resistant Bacteria in the News

To our knowledge, no one has taken a poll recently in which the public was asked to comment on the importance of antibiotics and the possible future impact of antibiotic-resistant bacteria. If such a poll were to be conducted, however, it would not be surprising to find that most members of the public are confused about antibiotics and antibiotic-resistant bacteria, if they think about them at all. A factor that has contributed in a major way to this confusion is the episodic way in which the press has covered the topic of antibiotic-resistant bacteria. Not only is news coverage sporadic and tending toward the sensational, but the antibiotic story constantly morphs into new forms.

There are at least three problems facing reporters who are assigned to cover the antibiotic resistance story. First, most reporters do not have much (or any) background in the basic biology of bacteria. If reporters are confused about whether the cause of the disease du jour is a virus or a bacterium, they are equally confused about what antibiotics are and about how to handle the sometimes conflicting views of scientists and physicians about whether there is actually an impending threat to the efficacy of antibiotics. Second, antibiotic use has become so pervasive in modern times and so varied in its application that it is sometimes difficult even for scientists and physicians to keep up with the most recent incarnation of the antibiotic resistance problem.

Third, and perhaps most serious, the antibiotic resistance story is rich in potential villains. Careless physicians who cavalierly dispense antibiotics like Halloween candy, profit-hungry pharmaceutical company CEOs who care only about the financial bottom line and not about public health, reckless farmers who misuse antibiotics in the hope of a miniscule profit advantage—so many possibilities for a harried reporter facing a deadline and the demands of an imperious editor! Never mind that these portrayals are basically inaccurate and ultimately destructive to rational public debate about how to manage the antibiotic discovery process or antibiotic use guidelines. Such stereotypes sell newspapers and magazines.

To be fair, it has been our experience that the vast majority of reporters, who may also have health problems and concerns about their families, want to get the story right. It's just that it's hard for them to get a handle on this constantly changing area. To appreciate their problems, let us take a look at the coverage of the antibiotic resistance story over the past couple of decades.

Reporters Discover a New Scare Story

Stories about antibiotic-resistant bacteria were rarely seen in the media prior to the 1990s. Starting in the mid-1990s, however, stories about antibiotic-resistant bacteria were suddenly everywhere. True to sensationalist form, or perhaps because of a lack of understanding of the problem, the press tried to turn what was actually a slowly developing insidious trend into a galloping crisis. Such sensational headlines as "Return to the Pre-Antibiotic Era" and "Superbugs on the March" conveyed the impression that doom was imminent. Writers of such stories asked, "What will happen if we lose antibiotics?" as if such a loss was just around the corner.

A popular theme of these stories was that we were in imminent danger of returning to the preantibiotic era, the bad old days when people died young from bacterial pneumonia, wound infections, and tuberculosis. In this new era, surgery and cancer chemotherapy would become too risky for any except the most desperate patients, and organ transplants would become a thing of the past. Even the normally staid scientific journal *Science* succumbed to this brand of hysteria. A 1996 cover of *Science* featured a diptych. On the left panel was a painting by the famous 15th century master Bruegel, a painting that was inspired by the bubonic plague. In the painting, piles of skeletons are being conveyed to wherever skeletons went in those days. On the right panel was a painting by a modern artist, who, though separated by centuries from Bruegel, had a similarly pessimistic outlook about human prospects. This painting depicts a lurid inner city landscape, lit by fires in the background and punctuated by an ambulance racing past. In the foreground, skeletons, with obviously sinister intentions, are consorting with the still living, some of whom are trapped in a skeleton's embrace. The message of this diptych was clear. A modern plague is coming soon, in the form of bacteria that are resistant to antibiotics, and you are all going to die.

A More Realistic Vision of the Future

If we lose antibiotics, will we return to a truly preantibiotic era? Before considering this question, let's define what we mean by *lose*. Although there are now some strains of bacteria that are panresistant, or resistant to virtually all available antibiotics, these strains are still in the minority. Most disease-causing bacteria remain susceptible to at least a few antibiotics. There even seem to be some bacteria, such as *Streptococcus pyogenes* (the cause of strep throat and a common cause of wound and bloodstream infections), *Chlamydia trachomatis* (the cause of a gonorrhea-like infection that can cause infertility and ectopic pregnancy), and *Treponema pallidum* (the cause of syphilis), that have remained steadfastly susceptible to most antibiotics. Scientists do not know why this is, and we may be in for some unpleasant surprises in cases like these if these bacterial slow learners finally catch up with the rest of the class, but for now it appears that some serious diseases would still be treatable.

Even in the case of bacterial species that are noted for resistance to antibiotics, there are some strains that have remained susceptible to most antibiotics. There may be an important lesson in this. Why would the incidence of resistant strains within a species rise to, say, 60% and then level off, leaving 40% of strains persistently susceptible? This is a phenomenon scientists don't understand, but such trends have been observed. Instead of focusing exclusively on strains that are becoming resistant to antibiotics, perhaps scientists ought to be paying some attention to those strains that seem not to have gotten onto the resistance bandwagon. In any event, there will probably not be a total loss of antibiotics. This is not to say that the situation will not be grim, but it may not become completely hopeless.

Another important factor to consider is that over the past 60 years since antibiotics first burst onto the clinical scene, we have learned a lot about preventing infections, knowledge that people in the preantibiotic era did not have. For example, our ability to prevent the transmission of infection in hospitals by using disinfectants and antiseptics and, in some cases, by isolating patients has improved considerably. Of course, nurses and physicians would need to be even more fastidious about cleansing their hands between patients and using latex gloves properly. Unfortunately, many hospital staff members cut short such hygienic practices because of complacency about the ability of these measures to control disease and the rushed schedules that are increasingly encountered by hospital personnel.

The introduction of alcohol-based disinfectant lotions that allow hospital staff members to disinfect their hands without having to find a sink is a potentially important development. Staff members simply carry a tube of lotion that they can apply as they walk from one place to another. Infection prevention and control in the community have also improved compared to the preantibiotic era. Moreover, improved nutrition and less crowded conditions have improved the robustness of the defenses against infection so that we do not get sick as readily.

Another advance that was not available in the preantibiotic era and that may help to save surgery from the worst effects of antibiotic-resistant bacteria is the advent of less invasive forms of surgery, such as surgery that requires only tiny cuts and laser surgery, which requires no breaching of the skin at all. These types of less-invasive surgeries, along with techniques that allow more rapid operations and thus reduce the opportunity for bacterial contamination of open surgical wounds, will help significantly to prevent postsurgical infections. Plastic implants that discourage bacterial growth are also an encouraging development.

On the cancer front, new "smart cancer drugs" that target the tumor cells and do not decimate the white cells that keep invading bacteria under control will lessen the risk of a cancer patient developing an overwhelming bacterial infection. Finally, scientists are exploring new ways to use vaccines and nonantibiotic strategies such as viruses that kill bacteria (bacteriophages) or bacterial proteins that kill bacteria (bacteriocins). Nonetheless, a world in which most antibiotics no longer work would be far grimmer than the world we have come to take for granted.

Would the Loss of a Cure Undermine Public Confidence in the Health Care System?

Now that the good news has been given, we turn reluctantly, but resolutely, to the bad news about a postantibiotic era. An aspect of antibiotic loss that is hardly ever discussed is the public's reaction to a situation in which cures that had been taken for granted suddenly become less effective or no longer available. In the preantibiotic era, people had not experienced the cures made possible by antibiotics, so their expectations were not very high. In a postantibiotic world, people would know what they had lost, and chances are that they would not be happy about it, to put it mildly.

The loss of an effective cure would be an historical first, and no one has any idea how the average person would respond to such a change in fortune. What would the loss of antibiotic cures do to public confidence in the medical profession? Would people still take a doctor's advice on procedures that are still effective, such as hygiene and vaccination? Surprisingly, futurists and psychologists have shown little interest in this question, the answer to which could have a massive impact on attitudes toward the health care system.

A science fiction author, Nancy Kress, has gone where futurists and psychologists have not dared to venture and has asked the question: How would people in a small city react to the loss of antibiotics? In her science fiction story "Evolution" (*Isaac Asimov's Science Fiction Magazine*, Oct. 1995) Ms. Kress takes a look at how the loss of all but one antibiotic, and that antibiotic beginning to fail, might affect the people in her fictional small city. Her view of what would happen is probably not far off the mark.

The story starts with the murder of a doctor by a vigilante group. The group has been formed by middle- and upper-income citizens. The doctor's crime was that he had treated a very sick child from a low-income family with the one remaining effective antibiotic. Knowing that antibiotic use engenders antibiotic resistance, the members of the vigilante group are sophisticated enough to conclude that they did not want unnecessary (by their definition) use of antibiotics on the "wrong people." This same vigilante group, it turns out, had previously blown up a bridge, making access to a nearby hospital, where antibiotics are available, more difficult. The story ends when someone, presumably a member of the vigilante group, blows up the hospital itself because of rumors that a bacterium resistant to the last antibiotic has appeared in that building and could spread to the community. This story is, of course, fiction, and we hope it will remain so, but it makes some important points that public health personnel would do well to consider in advance.

Antibiotic Resistance Issues Crop Up in Unexpected Places—Like Crops

Initially, news stories about antibiotic-resistant bacteria focused on resistance engendered by overuse of antibiotics by physicians. Soon, however, other aspects of the antibiotic resistance story, aspects connected to agriculture, began to surface. The first of these was connected to concerns about the safety of genetically modified plants as foods for humans and animals. From there, the antibiotic story grew to include agricultural use of antibiotics and antibiotic-resistant bacteria in the food supply. The discovery that antibiotics could be found in water supplies led to concerns about whether antibiotics could be acting as pollutants in some circumstances. Finally, the bioterrorism attack that started in October 2001, when an unknown person mailed letters laced with spores of *Bacillus anthracis*, started a panic about the safety of the mail system. The effective use of antibiotics such as ciprofloxacin (Cipro) and tetracycline to prevent or cure anthrax was reassuring, but people immediately began to wonder if antibiotic-resistant strains of *B. anthracis* would be the next bioterror threat. It seemed that there was no escape from antibiotic-resistant bacteria.

After visiting the possible Armageddon of a world in which antibiotics no longer work, the debate over the possible hazard posed by antibiotic resistance genes in genetically modified (GM) plants may seem frivolous. However, since this issue has been raised and covered ad nauseum in the press, it is worth reviewing.

At first glance, it might seem odd that the debate over whether GM plants are safe for human and animal consumption would have anything to do with antibiotic resistance genes, but oddly enough it does. During the genetic modification process used to construct the earliest GM plants, an antibiotic resistance gene, which had been used in the early cloning steps in bacteria, went along into the plant with the gene of interest, i.e., BT insecticidal toxin gene, and was incorporated into the plant genome.

Starting in 1995, critics of GM plants began to raise the question of whether DNA containing the antibiotic resistance gene could be released from the plant as plant tissue moved through the human or animal intestinal tract and enter the bacteria that are normally found in the colon, thus producing new antibiotic-resistant strains of bacteria. Many of the bacteria that normally reside in the intestinal tract of a healthy person can turn deadly if they escape the colon and enter the bloodstream, whether during surgery or some other trauma to the intestinal area. Increased resistance among these intestinal bacteria could increase the risk of not being able to treat such infections effectively.

In chapter 8, we will explain why scientists almost unanimously agree that this concern is not a serious one. Nonetheless, for several years, this controversy dominated the deliberations of the regulatory commissions assigned to assess possible risks of GM crops. A toxic side effect of this debate was that it diverted attention from real antibiotic resistance problems.

A good analogy is to imagine yourself sitting quietly in your kitchen when you suddenly hear loud noises on your front porch. You run to the window and see several thugs with a crowbar trying to break through your front door. You rush to your phone to call 911, but all you get is a recorded message. The message says that no police are available to answer your call because they are all attending a meeting to determine how they will respond if space aliens land in your town. The thugs on the porch represent, of course, the increasing death toll being taken by antibiotic-resistant bacteria that have arisen completely independently of GM crops and are killing thousands of people every year. The space aliens, in the form of the adverse health consequences of consuming GM crops, have yet to materialize.

Bioterrorism Comes to Town

An antibiotic story that hit the press in a big way and caused a lot of public discussion of the importance of antibiotics and the ways in which different antibiotics are used was the anthrax attack of 2001. In October 2001, letters laced with spores of *B. anthracis*, the cause of anthrax, first appeared and caused a widespread panic. The vehicle for dissemination of the spores, the postal system, was completely unexpected as a terrorist target. Like the food and water supply, the postal system touches the lives of virtually everyone in the country, so everyone felt close to the story.

Fortunately, scientists already knew a lot about anthrax and its treatment. Antibiotics given soon after exposure avert the most fatal form of anthrax, inhalation anthrax. An antibiotic called ciprofloxacin, or Cipro, was administered to workers in the congressional office buildings and news buildings where letters containing the spores had been opened. By contrast, an antibiotic called doxycycline was administered to postal workers who might have been exposed to the spore-laden letters.

Soon, people all over the United States were asking their physicians whether they should take Cipro. The course of Cipro therapy required taking Cipro daily for at least 60 days, an unusually long period of therapy for an antibiotic. This course was chosen because no one knew how long Cipro, which is a relatively new drug, would take to eliminate the bacteria. There had been no cases of human anthrax in the United States for many years, so there was no standard of reference. As it turned out, few people completed the full course of Cipro therapy because of the side effects (nausea, diarrhea). Nonetheless, the timely intervention of antibiotics prevented further deaths.

Many other antibiotics besides Cipro are effective against anthrax. One is doxycycline, an antibiotic that is a lot cheaper and has fewer side effects than Cipro, but many people felt, wrongly, that Cipro was the best choice. A heated debate over the relative merits of Cipro and doxycycline developed early in the anthrax scare period. One reason, already mentioned, was that congressional staff members and other elite groups were given Cipro, whereas postal workers were given the much less expensive antibiotic doxycycline. Reporters, postal workers, and their unions immediately scented a class warfare issue. To make matters worse, Cipro was, technically, the only drug approved by the U.S. Food and Drug Administration (FDA) for the treatment of anthrax. Doxycycline was not approved, in the sense that anthrax was not listed among the diseases for which doxycycline was recommended.

This case illustrates a problem in interpreting the meaning of "FDA approval" of a drug. There are many cases in which the FDA has not approved a drug for a certain application because it was ineffective or dangerous, but in the case of doxycycline, economics and timing of approval were responsible for the "approved" status of Cipro and the "not approved" status of doxycycline. Ciprofloxacin is a relatively new drug that is still under patent protection, hence, the high price. When the company that produces Cipro sought FDA approval for its drug, it included a long list of possible applications of the new antibiotic. Anthrax was included on this list. Probably anthrax was included, despite the fact that there had been no known human cases of inhalation anthrax in the United States for years, because there was talk about the possible use of *B. anthracis* as a bioweapon.

In earlier days, when doxycycline came up for FDA approval, no one in their right mind would have listed anthrax as a disease treatable by the antibiotic because anthrax was so rare and there was a strong treaty in force that banned the development and use of biological weapons. This was long before the first Gulf War, in which Saddam Hussein reportedly used spores of *B. anthracis* as a weapon. So, although infectious disease experts agreed that doxycycline would be a very effective treatment for anthrax, the fine print on the literature about the drug did not include this application. Even after there was talk of the possible use of *B. anthracis* as a bioterror weapon, it was hardly worthwhile for the manufacturers of doxycycline to get an FDA approval because doxycycline was no longer under patent protection.

Ironically, the supposedly favored congressional workers got a drug that was hideously expensive (initially about $700 for the course of therapy) and had unpleasant side effects, whereas the postal workers got an equally effective but much cheaper antibiotic with few side effects.

The Cipro experience raises another problem that has been little discussed. When a physician uses an antibiotic to treat a person for a specific infection, he or she also "treats" the bacterial populations that normally occupy the skin, mouth, intestine, and vaginal tract of the patient. Resistance to antibiotics in the ciprofloxacin family arises very readily when bacteria are exposed to these antibiotics, especially when the antibiotic concentrations in these various parts of the body are not high enough to kill the bacteria. Physicians had no choice but to treat people who might have been exposed to a life-threatening disease. Paradoxically, the treatment that undoubtedly saved some of them may have put the remainder at increased risk of future untreatable disease for reasons already mentioned in connection with the GM crop debate.

The bacteria that normally inhabit the human body, the so-called normal microbiota, are usually protective or neutral, but under some circumstances they can cause serious infections. Members of this normally innocuous bacterial population, for example, often cause postsurgical infections because they have the opportunity to enter areas of the body from which they are usually excluded. If those bacteria have become resistant to one class of our frontline antibiotics, the ciprofloxacin class of antibiotics, a valuable and perhaps life-saving class of antibiotics is no longer useful for treatment of postsurgical infections caused by them.

In addition to this, the physician evaluating the patient several years hence might not think to ask about previous use of ciprofloxacin and might thus assume that this antibiotic, which is normally a very effective antibiotic, would work. Loss of valuable time, engendered by the need to acknowledge failure of ciprofloxacin and to try some other antibiotic, could mean the difference between life and death for the patient. In this respect, doxycycline would have been a much better choice than Cipro for preventing anthrax, because resistance to doxycycline is already widespread in human intestinal bacteria and doxycycline is not often used for the treatment of possible multidrug-resistant postsurgical infections.

How're You Going to Keep Bacteria Down on the Farm, after They've Seen Antibiotics?

The old World War II song, "How're you going to keep them (meaning the soldiers, of course) down on the farm, after they've seen Paree" has an eerie resonance with the current debate about what happens to bacteria that become resistant to antibiotics because of antibiotic use in agriculture. By June 2002, there were two bills before Congress that would have drastically limited agricultural use of antibiotics—the Brown bill in the House and the Kennedy bill in the Senate. Not surprisingly, the lobbyists for the agricultural pharmaceutical companies and the groups that represent farmers came out in force to oppose these two bills. Also not surprisingly, neither bill passed.

In June 2003, a force that was far more powerful than Congress in the area of agricultural use of antibiotics entered on the side of limiting the use of antibiotics as growth promoters in animal husbandry. The fast-food giant McDonald's announced rather unexpectedly that henceforth, all of its meat suppliers were expected to reduce, then eliminate, the use of antibiotics as growth promoters. The public and the medical community may not listen to research scientists, but when McDonald's speaks, they are all ears.

Why have people become concerned about the agricultural use of antibiotics to the point that even environmental advocacy groups have taken up the issue? Partly, the answer boils down to the amount of antibiotic used. Although the precise figures are controversial, scientists at the Centers for Disease Control and Prevention and the Institute of Medicine estimate that at least as many tons of antibiotics are used in agriculture as are used to treat human disease.

Complacency about agricultural use of such large quantities of antibiotics persisted for a long time in part because the names of many of the agriculturally used antibiotics were different from those of antibiotics used to treat people. Thus, it appeared to the average person who decided to take an interest in agricultural antibiotic use that such use posed no danger to human medicine. However, many of the agriculturally used antibiotics are closely related in structure to important antibiotics used in humans and can cross-select for resistance to the antibiotics used in treatment of human infections (Table 4.1).

An example is avoparcin, an antibiotic used as a growth promoter in Europe until it was banned in 1997. Avoparcin is a structural analog of vancomycin, one of the last-ditch human use antibiotics, so avoparcin can select for bacteria that are also resistant to vancomycin. The sheer volume of use in agriculture is enough to get some people worked up. Add to this the fact that most antibiotics do not magically disappear after they have passed through an animal, and you have the specter of tons of

Table 4.1   Antibiotics used in agriculture that cross-select for antibiotics used in human medicine

Agricultural antibiotic	Analogous human use antibiotic to which agricultural use antibiotic selects resistance	Use of antibiotic in human medicine
Avoparcin	Vancomycin	Postsurgical infection, bacterial pneumonia
Tylosin	Erythromycin	Sexually transmitted diseases, postsurgical infections, lung infections
	Synercid	A newly introduced antibiotic for treating multidrug resistant, postsurgical, and other serious infections
Fluphenicol	Chloramphenicol	Not as widely used as vancomycin and erythromycin because of side effects, but coming back into favor for some diseases

antibiotics being washed into water supplies, with unknown consequences.

A more immediate cause for concern in the view of scientists is that the antibiotic-resistant bacteria selected on the farm enter and move through the food supply into the human body. Why this should be a concern will be covered in more detail in chapter 8. For now, it will suffice to point out that if a person acquires a bacterial population in his or her intestine that contains high concentrations of bacteria resistant to antibiotics, that person is at greater risk for postsurgical infections that are difficult to treat. The magnitude of this added risk is controversial, but few would like to do the experiment of colonizing a substantial portion of the human population with antibiotic-resistant intestinal bacteria to find the answer.

Antibiotics are used for three purposes in agriculture. One is to treat sick animals—an application with which few would quarrel. This application accounts for at most 10 to 15% of the antibiotics used in agriculture. A second use, which accounts for at least 30% of antibiotic use, is prophylaxis, the administration of antibiotics to prevent disease. This use is of importance on factory farms because infectious diseases can spread like wildfire through crowded populations of animals. Animals that are bred for maximal production of meat, eggs, or milk can be somewhat immune-compromised because protein is being diverted from the immune system—a major consumer of protein in a normal animal. Hence, the need to prevent disease is even more urgent.

The most controversial agricultural use of antibiotics is as growth promoters (Table 4.2). That is, some antibiotics seem to give some animals a growth advantage. The animals do not necessarily grow larger, but they gain weight more rapidly. Given the slim profit margins most farmers face, even a 4 to 5% increase in weight gain (the level of effect of the best growth-promoting antibiotic) can be critical. Figures quoted for the amount of antibiotic used for growth promotion vary from as low as 15% to as high as 50%. The variation in estimates of growth promotional use arises from how you define "prophylaxis" and "growth promotion." Some scientists have suggested that antibiotics that have been called growth promoters actually work by preventing disease (as prophylaxis), a type of use that is more acceptable to the public than growth promotion.

The source of the problem with defining growth promotion is uncertainty about how growth promoting antibiotics work. Although industry scientists point to published papers on this subject, these papers merely describe changes in the animal's physiology, not the mechanistic basis for these changes. The possibility that some growth-promoting antibiotics may actually work because they prevent disease, despite the very low

Table 4.2   Examples of antibiotics used in animal agriculture and human medicine

Antibiotic class	Animal use			Human use
	Treatment	Prophylaxis	Growth promotion	
Aminoglycosides (e.g., gentamicin)	Yes	Yes	No	Yes
β-Lactams				
Penicillins	Yes	Yes	Yes	Yes
Cephalosporins	Yes	Yes	No	Yes
Macrolides (e.g., erythromycin)	Yes	Yes	Yes	Yes
Fluoroquinolones	Yes	Yes	No	Yes
Sulfonamides	Yes	No	Yes	Yes
Tetracyclines	Yes	Yes	Yes	Yes

levels of antibiotic used for growth promotion, has already been mentioned.

Another possibility is that antibiotics that promote growth have an effect on bacteria, or even a direct effect on the animal, that leads to reduced turnover and excretion of the cells that line the intestine. The constant replacement of intestinal cells is an important defense of the body against bacterial infections because it helps to prevent bacteria that have bound to intestinal cells from invading these cells and moving further into the body. This constant replacement, however, consumes a lot of carbon and energy that could have gone to increase muscle and fat mass. Thus, any treatment that slows the turnover of intestinal cells could significantly increase weight gain.

Arguments about whether an antibiotic is being used prophylactically or as a growth promoter occur because the answer has practical consequences. In both the Brown and Kennedy bills, what was to be forbidden was "nontherapeutic use" of a number of antibiotics. "Nontherapeutic" is in the eye of the beholder. Is the use of antibiotics as growth promoters nontherapeutic? The answer to this depends on your version of how antibiotics promote growth. If antibiotics promote growth by reducing the animal's load of certain bacteria, then such antibiotics could be considered therapeutic. Is prophylaxis nontherapeutic? Some would argue that treatment of a herd or flock in which there are currently no sick animals would fit into this category. Others, mindful of the potential for devastating disease in these crowded animal populations, would disagree.

They're Everywhere! They're Everywhere! Antibiotics as a Broader Environmental Issue

Just when scientists thought antibiotic-related controversies could not get any more complex, they got another rude awakening. In 1998, some German scientists were measuring concentrations of various chemicals in water that had just been released from a sewage treatment plant. They were using chromatography to separate and identify compounds in the water. To their surprise, they noticed an unexpected peak in the chromatogram and proceeded to identify it as a fluoroquinolone antibiotic. This should not have been a surprise, because some antibiotics are quite stable in the environment, but people were surprised because they had gotten used to the idea—which has no scientific basis whatsoever—that antibiotics vanished magically when flushed down the toilets of homes and hospitals or leached into groundwater from piles of manure on farms.

Scientists now have to adjust to the fact that antibiotics are flowing from sewage treatment plants and leaching from manure storage pits on farms into the Earth's water supply. Antibiotics are not the only chemicals making their way into the water supply. Heart medications and hormones are also being found in surprisingly high concentrations.

What are the consequences, if any, of turning water that may be used for drinking or water that flows into nature preserves into a witches' brew of chemicals, including antibiotics? Scientists are investigating the consequences of this type of pollution more actively and are beginning to suspect that antibiotics and other medicinal compounds could be having an effect on microbes, plants, and perhaps even animals. The magnitude of this effect remains to be determined.

Issues to Ponder

1. It is obvious why reporters sensationalize stories about antibiotic-resistant bacteria. First, scaring people grabs their attention. Second, the average reporter does not have the time to learn enough about antibiotic-resistant bacteria to delve below the surface. What would it take to make you read a news article or opinion piece if it was not a sensational scare piece? Economic impact? Specific effects on your family?

2. How would you cover the story of antibiotic uses in agriculture in a way that is fair to farmers and to the public at large? To what extent would you take into consideration the farmers' concerns about the possible adverse economic consequences to them of ceasing most agricultural use of antibiotics?

3. There are some lawyers who are beginning to specialize in suing doctors and hospitals over hospital-acquired infections, especially those caused by antibiotic-resistant bacteria. There has been little press coverage of these suits. Should they receive more attention? As a reporter, how would you cover such a story?

5

Antibiotics That Inhibit Bacterial Cell Wall Synthesis

Penicillin and related antibiotics such as ampicillin or amoxicillin are truly "cradle-to-the-grave" drugs. Small child with an earache? The doctor will likely prescribe a penicillin. A young woman with a urinary tract infection? Ampicillin is an old standby for treatment. An elderly man with a case of bacterial pneumonia? You guessed it! A penicillin derivative. In this chapter, we explain how penicillin and structurally related antibiotics kill bacteria. We also introduce an antibiotic called vancomycin that has become an important frontline drug for treating bacterial infections that are resistant to other antibiotics.

Structures of antibiotics mentioned in this chapter are shown in appendix 1.

The Rigid Bacterial Cell Wall—Essential Armor Coating for Bacteria

Most bacteria are covered by a rigid cell wall that not only defines their shape, the bacterial equivalent of a fashion statement, but also maintains their structural integrity. This rigid coat is necessary because most bacteria live in liquids whose concentration of salts and other small molecules is much lower than the concentration of these same small molecules inside the bacterial cell. Since water can flow across the bacterial membrane, it would rush into the cell, causing the cell to swell and burst, if such swelling were not prevented by a tough cell wall that covers the surface of the bacterium. An antibiotic that impairs the strength of this cell wall will kill the bacterium by causing the bacterium to burst. It is just as hard today as it was in Victorian England to put Humpty Dumpty back together again.

The main component of a bacterial cell wall is a complex mesh-like polymer called peptidoglycan, which gives the cell wall its strength. There are two main types of bacterial cell wall. Originally this difference was noted because of a staining procedure that differentiates bacteria into two categories, gram positive and gram negative. Hans Christian Gram, the inventor of this staining procedure, developed it long before anything was known about the basis for the staining difference. Even before the basis of the difference in staining was understood, the procedure became widely used in clinical laboratories as a simple, early step in the characterization of bacteria in a clinical specimen.

In the Gram stain procedure, bacteria are first attached to a glass slide by heating the slide. Then a combination of two compounds, the blue stain crystal violet and iodine, is applied to the bacteria on the slide. The bacteria take up the stains, making them a dark blue color. The bacteria on the slide are then washed gently with an acetone-alcohol mixture. This wash removes the blue dye from some types of bacteria, rendering them colorless under the microscope, whereas other bacteria retain their blue stain. Since the bacteria that lose the blue stain during the acetone-alcohol wash become invisible, they have to be stained with a second dye, safranin, which gives them a light red color. The bacteria that retain the blue dye also stain with safranin, but the red dye makes the blue appear deeper and more purplish. Bacteria that stain blue are called gram positive and bacteria that stain red are called gram negative.

Simplified diagrams of the cell walls of gram-positive and gram-negative bacteria are shown in Fig. 5.1. Peptidoglycan is present in both cases, but the peptidoglycan layers differ in thickness, being much thicker in the gram-positive cell wall than in the gram-negative cell wall. Peptidoglycan looks a lot like a chain link fence. It consists of a linear backbone composed of sugars (glycans). The glycan strands are cross-linked with chains of amino acids (peptides). A more detailed diagram of the structure of peptidoglycan is shown in Fig. 5.2. The antibiotics penicillin and vancomycin undermine the stability of this very strong meshwork structure by preventing the peptide cross-links from forming.

Other antibiotics, such as fosfomycin and bacitracin (a common ingredient in over-the-counter antibiotic ointments), act at earlier stages in peptidoglycan synthesis but ultimately have the same effect of preventing the peptidoglycan structure from forming. Peptidoglycan is such a large structure that it could not be constructed in the bacterial cytoplasm and then exported intact to the bacterial surface through the bacterial cytoplasmic membrane. Instead, bacteria construct the sugar-peptide subunits

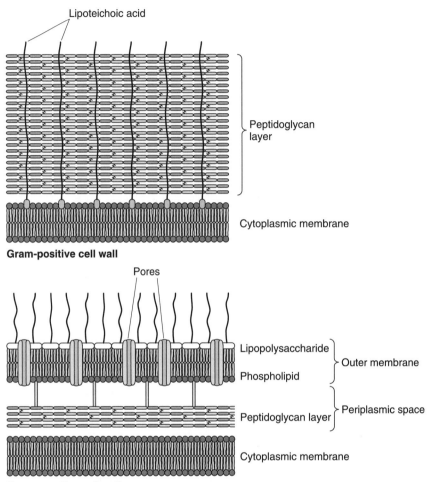

Figure 5.1 Comparison of the structures of gram-positive and gram-negative bacteria. The cytoplasm of both types of cells is surrounded by a cytoplasmic membrane. Outside the cytoplasmic membrane is the complex mesh-like polymer, peptidoglycan. The peptidoglycan in the gram-positive cell wall consists of many layers. Other substances such as lipoteichoic acid are woven through the peptidoglycan. Gram-negative cells have a thin layer of peptidoglycan, which is covered by an outer membrane. The outer membrane consists of one layer of phospholipids and one layer of lipopolysaccharide.

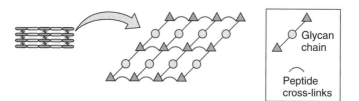

Figure 5.2   The structure of peptidoglycan. Peptidoglycan consists of chains of glycans (sugars) that are linked together with chains of amino acids (peptides).

of peptidoglycan in their cytoplasm. These sugar-peptide units are then exported through the cytoplasmic membrane and assembled on the surface of the membrane into peptidoglycan, much as bricks can be carried out of a house to construct a wall outside. A special lipid carrier, called bactoprenol, carries out the export process. The antibiotic fosfomycin prevents the synthesis of the peptidoglycan subunit, and bacitracin interferes with its export by interfering with the functioning of bactoprenol. Penicillin and vancomycin prevent the last step in peptidoglycan synthesis, the cross-linking that makes peptidoglycan so strong.

Bacteria that stain gram positive have a cell wall that consists of many layers of peptidoglycan. The surface of this peptidoglycan layer is covered with proteins that perform various functions such as allowing the bacterium to bind to human tissues but they do not provide much of a barrier to small molecules. Antibiotics can diffuse readily through this porous structure, however. This is important because the proteins targeted by antibiotics like penicillin and vancomycin, the proteins that assemble the exported units into the peptidoglycan mesh, are located in the cytoplasmic membrane and are exposed on the membrane surface. Thus, penicillin and vancomycin need not cross the cytoplasmic membrane and enter the cytoplasm in order to work, they need only reach the membrane surface. This is not true for fosfomycin, which has to enter the cytoplasm of the bacterial cell in order to prevent the synthesis of the sugar-peptide subunit.

Bacteria that stain gram negative have a more complex cell wall. Their peptidoglycan covering is only a few layers thick, but it is stabilized and strengthened by a second membrane, the outer membrane (Fig. 5.1). The outer membrane is an unusual membrane. Unlike most biological membranes, it does not consist of two layers of phospholipids (phospholipid bilayer), but consists instead of one layer of phospholipids and one layer of a substance called lipopolysaccharide. Lipopolysaccharide has a lipid component that is embedded in the outer membrane and a polysaccharide portion that sticks out from the surface of the bacteria.

The outer membrane would keep out nutrients and other essential compounds needed by the bacteria if it were not for some open channels in the outer membrane (pores) that allow small molecules to diffuse into the space between the outer membrane and the cytoplasmic membrane. This zone is called the periplasmic space. The proteins that form the pores in the outer membrane are called porins. The openings in these pores prevent some large antibiotics from passing through the outer membrane, making gram-negative bacteria more resistant to certain antibiotics than gram-positive bacteria. Also, bacteria can modify the porin proteins so that the pores become smaller in diameter. In this case, even smaller antibiotics may not diffuse as readily through the outer membrane pores as they did before and thus do not easily reach the surface of the cytoplasmic membrane where the process of peptidoglycan construction occurs.

Penicillins and Cephalosporins

Penicillins and a related type of antibiotic called cephalosporins have long been mainstays of antibiotic therapy. Some members of the penicillin family are shown in Fig. 5.3 and Appendix 1. Although these molecules look at first as if they are very different in structure, they have in common a four-membered ring called the β-lactam ring, which is circled in Fig. 5.3. This four-membered ring consists of three carbon atoms and one nitrogen atom (the β-lactam ring) and is the active portion of penicillins and cephalosporins. This conserved ring is the reason why members of the penicillin-cephalosporin family are called β-lactam antibiotics.

Recently, two more types of β-lactam antibiotics have been added to the β-lactam family: carbapenems and monobactams. The main motivation for developing these new β-lactam antibiotics was to counter increasing resistance of some bacteria to penicillins and cephalosporins. Monobactams also have had an additional benefit. Some people become allergic to penicillin and most related β-lactam antibiotics. This happens because the antibiotic binds covalently to a blood protein. The body now sees this β-lactam-protein hybrid as foreign protein and the immune system mounts a response to it. The result is that, in the future, the body's immune system attacks penicillin and other β-lactam antibiotics whether they are

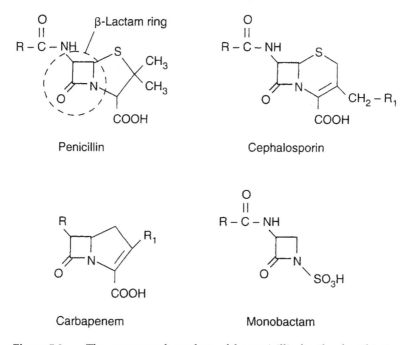

Figure 5.3   The structure of members of the penicillin family of antibiotics (β-lactams). This family of antibiotics is characterized by a four-membered ring (the β-lactam ring).

linked to a protein or not. This attack triggers an allergic response that can be very serious. People who have become allergic to penicillin can safely receive monobactams because these β-lactam antibiotics do not trigger the same allergic response as other members of the β-lactam family.

How Penicillin Kills Bacteria

The β-lactam antibiotics kill bacteria because they undermine the structure of peptidoglycan sufficiently to cause the bacteria to explode due to internal pressures no longer countered by the peptidoglycan "girdle." How do the β-lactam antibiotics undermine the rigid structure of peptidoglycan?

The strength of the peptidoglycan meshwork lies in its peptide cross-links. Without those cross-links, the glycan strands are simply wrapped about the bacterial surface like yarn on a ball. As you know if you have ever toyed with a ball of yarn, the strands of yarn are easily separated from each other, even while they are still wrapped around the ball of yarn. The weakened cell wall that lacks the stabilizing cross-links cannot contain the internal pressure of the bacteria, and the bacteria burst.

The cytoplasmic membrane of a bacterium is studded with an array of proteins that are exposed on the surface of this membrane. Some of these cytoplasmic proteins join the sugar molecules of the exported peptidoglycan subunit to make the glycan backbone. Others take the peptide components of the units and join them covalently with a peptide from an adjacent glycan strand, creating the final cross-linked structure of peptidoglycan.

The β-lactam antibiotic binds covalently to these cross-linking enzymes and inactivates them. For this reason, these enzymes have been called penicillin-binding proteins. Since an important effect of penicillin is to prevent the formation of the cross-links that stabilize the peptidoglycan structure, then peptidoglycan newly synthesized in the presence of a β-lactam antibiotic lacks the strength of the normal β-lactam meshwork.

The β-lactam antibiotics exert their effect on the penicillin-binding proteins because the proteins doing the cross-linking first have to cleave the peptide bond between two D-alanine residues in the peptide to create the cross-link (Fig. 5.4). The β-lactam antibiotics are structurally similar to a D-alanine-D-alanine peptide and are thus attacked by the cross-linking enzymes just as if they were peptide amino acids. Unfortunately, the antibiotics are different enough from the D-alanine-D-alanine of the future peptide cross-link that the enzyme cannot transfer them to an adjacent

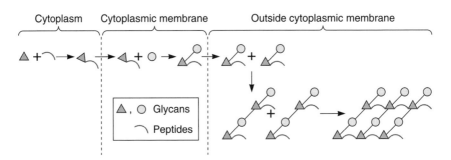

Figure 5.4   Cross-linking occurring during the synthesis of peptidoglycan. Sugar-peptide units are formed in the cytoplasm of the cell. These units are then exported across the cytoplasmic membrane to the exterior where they are assembled into peptidoglycan.

peptide. In trying to cleave the β-lactam antibiotic, the enzyme gets stuck in the process and the antibiotic gets stuck to the enzyme. This prevents the enzyme from further cross-linking the peptidoglycan.

The problem with the explanation just given, an explanation found in all microbiology textbooks, is that it doesn't really explain why preventing cross-link formation on newly synthesized peptidoglycan, leaving pre-existing peptidoglycan alone, leads to the near total breakdown of the peptidoglycan layer that is actually seen in the case of penicillin-treated bacteria. Only recently have microbiologists begun to revisit the question of whether β-lactam antibiotics have other effects on bacterial cells besides inhibiting the cross-linking reaction.

A part of the answer lies in a set of bacterial enzymes called autolysins. Autolysins are bacterial cytoplasmic membrane proteins that break down peptidoglycan. They are necessary because if a bacterium that is encased in a rigid cell wall needs to lengthen itself in order to divide, it needs some breathing room provided by a temporary breach in the integrity of the rigid peptidoglycan cell wall. Such breaches are potentially dangerous to the bacterium and thus are tightly controlled. How autolysins are activated and deactivated to allow bacterial division without causing the bacterium to explode is still not understood very well, but the answer to the larger question of how β-lactams kill bacteria is now believed to lie in the effects of these antibiotics on the activity of autolysins as well as in their direct action on the penicillin-binding proteins.

How Bacteria Become Resistant to β-Lactam Antibiotics

The first account of a bacterial strain that was resistant to penicillin was published at about the same time penicillin was introduced into medical use, but it took three decades for scientists to begin to understand how bacteria were protecting themselves against this powerful new antibiotic. The first mechanism of resistance to penicillin to be described was an enzyme called β-lactamase. This enzyme, as the name suggests, cleaves the β-lactam ring of penicillin, rendering the antibiotic inactive.

β-Lactamases seem to have evolved from the proteins that catalyze the cross-linking of peptidoglycan—the proteins that are the target of penicillin. When penicillin binds one of these proteins, the protein mistakes penicillin for the two alanines found at the end of the peptide that will normally participate in the cross-linking reaction. The cross-linking enzyme can start but not complete the hydrolysis of the β-lactam ring.

The result is that the partially hydrolyzed antibiotic is trapped in the active site of the enzyme and the enzyme no longer functions. A β-lactamase is a protein that has evolved in its structure to the point where it is able to complete the reaction, releasing a penicillin molecule with a broken β-lactam ring.

Since the target of penicillin is a set of proteins that protrude from the surface of the bacterial cytoplasmic membrane, bacteria must secrete their β-lactamases through the cytoplasmic membrane to the site where these penicillin-degrading enzymes can dispose of penicillin before the penicillin molecules have a chance to bind to and inactivate their target. In the case of gram-positive bacteria, the surface of the cytoplasmic membrane is the outside of the cell (Fig. 5.1), so the β-lactamase can diffuse away from the cell.

In the case of gram-negative bacteria, the β-lactamase is trapped between the cytoplasmic membrane and the outer membrane in the periplasmic space. Although low-molecular-weight components from the external environment, including penicillin and other β-lactam antibiotics, can diffuse through the pores in the outer membrane, the β-lactamase itself is far too large to do so. The β-lactamase functions somewhat like a guard dog that is confined to a yard by the fence provided by the outer membrane.

Gram-negative bacteria use their outer membranes in another way. They mutate the proteins that make up the outer membrane pores so that the opening becomes narrower. This change somewhat restricts the ability of penicillin and similar molecules to diffuse into the periplasmic space, thus making it easier for the periplasmic β-lactamase to inactivate all of the penicillin molecules. Mutant porins can be present without the β-lactamases but are not nearly as effective in protecting the bacteria from β-lactam antibiotics as they are when they are paired with the inactivating enzymes. In this combination, the mutant porins make it slow work for antibiotic outsiders to gain access to the yard, giving the β-lactamase "dog" a better chance to attack them as they try to enter.

Scientists Strike Back: Combating β-Lactamases

A bacterium produces a β-lactamase. Score one for the bacteria, but scientists have some tricks up their sleeves too. One way scientists have acted to combat bacterial β-lactamases is to develop modified forms of penicillin that are no longer attacked effectively by β-lactamases. Note that in

Fig. 5.3, the β-lactam ring is surrounded by other chemical groups. Some of these groups participate in binding the antibiotic to the β-lactamase. If they are altered, the ability of the β-lactamase to bind to and inactivate the antibiotic is decreased. These new antibiotics have been called β-lactamase-resistant β-lactams.

Unfortunately, as rapidly as scientists produce new forms of β-lactamase-resistant β-lactam antibiotics, the β-lactamases themselves mutate so that they become able to degrade these new antibiotics. These enzymes are called extended-spectrum β-lactamases (ESBLs) because they now have an extended spectrum of new β-lactam victims.

A second way to deal with bacterial β-lactamases is to develop an antibiotic preparation that contains both the antibiotic and an inhibitor of the β-lactamase that attacks it. A widely used example of this type of preparation is a combination of the β-lactam antibiotic amoxicillin and a β-lactamase inhibitor, clavulanic acid. This mixture is marketed under the trade name Augmentin. Clavulanic acid (Fig. 5.5) looks something like a β-lactam antibiotic but it does not inactivate the cell-wall-synthesizing enzymes. That is, clavulanic acid has no antibacterial activity. Its sole role is to bind to and inactivate the β-lactamases that are trying to destroy amoxicillin. Clavulanic acid molecules, like members of a football defense line, keep the β-lactamase out of action so that amoxicillin can make a "touchdown" to kill the bacterial cell.

Bacteria Develop Another Resistance Strategy That Trumps the β-Lactamase Inhibitors

The early successes of the β-lactamase inhibitor strategy made scientists optimistic that they were finally winning the fight against bacteria that were becoming resistant to β-lactam antibiotics. All they had to do was keep improving the ability of the β-lactamase inhibitors to inactivate the ever-evolving bacterial β-lactamases. In this way, old standbys like amoxi-

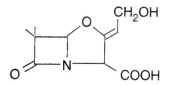

Figure 5.5   The structure of clavulanic acid.

cillin and ampicillin could be recycled. The gram-positive bacteria, and subsequently the gram-negative bacteria, had an unpleasant surprise in store for these scientists, however.

As already mentioned, the targets of the β-lactam antibiotics are the penicillin-binding proteins (PBPs), which catalyze the cross-linking of peptidoglycan. Scientists began to see bacteria with mutant PBPs that no longer bound β-lactam antibiotics. Uh oh. This resistance mechanism is not countered by β-lactamase inhibitors, because no β-lactamase is involved in resistance. A bacterium can, of course, combine mutant PBPs and β-lactamase for a very effective one-two punch, but at best the β-lactamase inhibitors will only counter one arm of the bacterium's resistance armamentarium. Mutant PBPs are now being seen both in gram-negative bacteria and gram-positive bacteria, so this strategy seems to be a "one size fits all" resistance mechanism. Score one more for the bacteria.

Rediscovering Another Type of Cell Wall-Synthesis Inhibitor: Vancomycin

The success of the β-lactam antibiotics, both in terms of efficacy and safety, was so impressive that little attention was paid for a long time to another antibiotic that also inhibited bacterial cell wall synthesis: vancomycin. Vancomycin and a related antibiotic called teicoplanin had been discovered during the 1960s. Vancomycin had two major drawbacks. First, the initial preparations had a brown color, a trait that caused some scientists to call vancomycin "Mississippi mud"—not much more appealing than the "earwax" described in chapter 2. The brown color was a tip-off that this vancomycin preparation contained a number of contaminants that would make it not only less effective but also possibly less safe.

A second drawback was that vancomycin was only effective against gram-positive bacteria. The reason for this is evident from looking at its structure (see appendix 1). Vancomycin is a big, rangy molecule that cannot diffuse through the outer membrane porins of gram-negative bacteria. When vancomycin was first discovered, physicians much preferred antibiotics that were effective against all bacteria, gram positive as well as gram negative. This allowed the time-consuming process of bacterial identification to be dispensed with and made it possible, in theory, to use a single antibiotic for many infectious disease applications. Such broad-spectrum antibiotics are particularly important in the early critical period when a physician suspects that the patient has a bacterial infection but has no idea of its species identification.

Another development that made vancomycin unappealing was that during the 1960s and 1970s, the gram-positive bacteria, which had dominated the disease picture in the 1940–1950 period, decreased rather dramatically in the number of cases they caused. The gram-negative bacteria such as *Escherichia coli*, *Klebsiella* species, and *Pseudomonas* species became the bacteria to beat in hospitals and clinics. In the 1980s, however, the gram-positive bacteria such as staphylococci and streptococci began to roar back into a position of prominence, for reasons that are not completely clear. A likely explanation is that the use of antibiotics that were most effective against gram-negative bacteria allowed the gram-positive bacteria to reclaim their starring role in the infectious diseases.

Vancomycin, which had been collecting dust on the shelf prior to the 1980s, began to look pretty good after all. A purer preparation that had lost the contaminants responsible for the brown color of the earlier preparations was soon available, and vancomycin soon became a front-line antibiotic for treating those gram-positive bacteria that had become resistant to other antibiotics.

Today, one of the hottest growing areas of antibiotic discovery is finding more antibiotics that are effective against the gram-positive bacteria. Ignoring the gram-negative bacteria, however, is as unwise today as was ignoring the gram-positive bacteria during the 1960–1970 period, because it is now clear that the spectrum of disease-causing bacteria can alter. Have scientists, physicians, and pharmaceutical companies learned the lesson that the spectrum of disease-causing bacteria is dynamic? Stay tuned, but don't get your hopes up.

How Vancomycin Works and How Bacteria Are Becoming Resistant to It

Vancomycin, like penicillin, inhibits the cross-linking reaction that finishes the construction of the peptidoglycan cell wall, but it does so in a different way. Whereas penicillin binds to and inactivates the enzymes that make the cross-links, vancomycin binds to the peptides that are slated to become part of the cross-linked peptidoglycan structure. Enzymes have a very precise lock-and-key interaction with their substrates, and anything that interferes with that interaction can stop the enzymes in their tracks. Vancomycin is a large, bulky molecule that binds to the part of the peptide the cross-linking enzymes would normally interact with and thus physically prevents the enzymes from carrying out the cross-linking reaction.

How could a bacterium become resistant to such an antibiotic? One would have thought that vancomycin would have been the ultimate resistance-proof antibiotic because becoming resistant to it would require the bacteria to change their cross-link peptide structure, a change that not only might confuse the cross-linking enzymes but would require a new pathway for synthesis of the peptide. Also, the bacterium would have to get rid of the old peptide or it would remain susceptible to vancomycin's action.

Unfortunately for us, some bacteria have managed to accomplish this feat. Virtually all of the resistance strategies described in this chapter and in subsequent chapters involve a single resistance protein, an enzyme that inactivates the antibiotic or an altered antibiotic target that is no longer affected by the antibiotic. Resistance to vancomycin actually involves several genes encoding several proteins that comprise a pathway for changing the peptidoglycan cross-linking peptides into a form that no longer binds vancomycin but will still be cross-linked by bacterial enzymes. There is also a gene that encodes an enzyme that degrades the original terminal D-alanine-D-alanine part of the cross-linking peptide (Fig. 5.6).

Fortunately, the most dangerous gram-positive pathogens, such as *Staphylococcus aureus* and *Streptococcus pneumoniae*, are currently susceptible to vancomycin, although resistance to vancomycin is beginning to appear in members of these species. The most feared development is the emergence of strains of *S. aureus* that started out being resistant to many antibiotics, including β-lactam antibiotics, and have now become resistant to vancomycin. The first reports of such bacterial strains have begun to appear. They have been called VRSA for vancomycin-resistant *S. aureus*. Fortunately, the few VRSA strains isolated so far have proven to be susceptible to at least one other antibiotic, but since *S. aureus* has shown an impressive adeptness at acquiring resistance to many antibiotics, this situation may not last very long. Earlier in this chapter, we mentioned that scientists began to see bacteria that were resistant to penicillin shortly after penicillin was introduced. Guess the species. You got it: *S. aureus*.

There may be some gram-positive bacteria that are naturally resistant to vancomycin because they have peptides in their peptidoglycan that do not have a terminal D-alanine-D-alanine peptide. This possibility has come up in connection with probiotics. Probiotics are preparations of presumably beneficial bacteria that are consumed daily in powder form or in

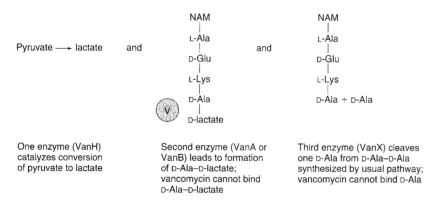

Figure 5.6   The mechanism of vancomycin resistance. Two enzymes in the resistance pathway replace the D-alanine-D-alanine with D-alanine-D-lactate, which does not bind vancomycin. One enzyme links D-lactate to D-alanine. A second one converts pyruvate, an intermediate in glycolysis, to D-lactate. A third enzyme hydrolyzes any normal D-alanine-D-alanine peptides that might form and be bound by vancomycin. The cross-linking enzymes of the cell seem to be able to handle D-alanine-D-lactate and thus do not need to be modified. The encircled V represents vancomycin, which is no longer able to bind. (Reprinted from A. A. Salyers and D. D. Whitt, *in Bacterial Pathogenesis: a Molecular Approach*, ASM Press, Washington, D.C., 2002.)

unpasteurized yogurt. Most of the bacteria in these preparations are *Lactobacillus* species. There is considerable controversy as to whether they have any scientifically provable health benefits, but the contention has always been that at least they do not harm. Imagine then, the shock to some of us in the resistance field to learn that most of the *Lactobacillus* strains being used in probiotics were resistant to vancomycin. Fortunately, this resistance seems to be an innate metabolic trait, a naturally occurring difference from other bacteria in the peptidoglycan peptide cross-link, and not a characteristic that can be transmitted to other bacteria.

What will replace vancomycin and its relatives? Unfortunately the list is not a long one. Some of the recent candidates such as the protein synthesis inhibitor Synercid have had an initially disappointing performance. Given that so many pharmaceutical companies have abandoned the search for new antibiotics for economic reasons, the list of possible vancomycin replacements may become a very short one indeed. This fact underscores the importance of protecting vancomycin and more recent derivatives of this antibiotic from overuse by physicians in the hope that the lifetime of these now-vital antibiotics can be prolonged until some substitute is found. This is going to be hard to do because surgeons, worried about postsurgical infections caused by bacteria that are resistant to other antibiotics, have increasingly tended to favor vancomycin to prevent and treat gram-positive infections. This type of decision increases further the overuse of vancomycin and provides a selective pressure for the emergence and dominance of vancomycin-resistant strains of *S. aureus* and other gram-positive bacteria.

It is difficult to do effective finger pointing at physicians who may have a bit of an itchy finger on the vancomycin trigger when you learn that an analog of vancomycin, avoparcin, has been used as a livestock feed additive in Europe (chapter 8). Avoparcin has a different name from vancomycin, but it cross-selects for resistance to vancomycin. Fortunately, the use of avoparcin in agriculture has been stopped, at least in developed countries, although there are persistent rumors that avoparcin use in agriculture continues in developing countries. The avoparcin example illustrates quite dramatically the importance of constant vigilance to protect front-line antibiotics from abuse in settings other than hospitals.

Issues to Ponder

1. Could a strategy in which two antibiotics that target the same step in peptidoglycan synthesis, such as a combination of a β-lactam antibiotic and vancomycin, work to counter bacterial resistance? If so, what would be its limitations? Can you guess why pharmaceutical companies have been reluctant to develop such preparations?

2. Antibiotics that prevent the synthesis of peptidoglycan are bactericidal. That is, they kill bacteria outright. In subsequent chapters, other antibiotics will be introduced that only inhibit the growth of bacteria, not kill them. These antibiotics are called bacteriostatic. Why do bacteriostatic antibiotics work in most people? For what types of patients might bactericidal antibiotics be essential? Keep in mind that bacteriostatic antibiotics rely on the immune system to clear up the bacteria.

3. The avoparcin debacle occurred in Europe at the same time as the intense controversy over the safety of genetically modified crops. Europeans made it clear that they wanted to be super-safe when it came to the food supply. Yet they clearly applied a different standard to antibiotic

resistance genes that had ended up in the plant cells as a consequence of cloning than to the agricultural use of avoparcin. How could such a situation develop? This is not a question with an agenda. We think it is important to understand why the same public could react in such different ways to the problem of antibiotic-resistant bacteria. We are, frankly, at a loss to explain this European conundrum. Any help from our reading public would be welcome.

6

Antibiotics That Inhibit the Synthesis of Bacterial Proteins

If your teenage child goes to a dermatologist for help in controlling acne, the dermatologist is likely to prescribe oral tetracycline. He or she may also prescribe a cream containing the antibiotic clindamycin (Cleocin). The similarity between tetracycline and clindamycin is more than skin deep. Although these two antibiotics are structurally different (structure wonks should go immediately to appendix 1), they have the same effect on bacteria. They bind to bacterial ribosomes, the tiny protein factories that translate messenger RNA (mRNA) into proteins, and prevent bacteria from producing essential proteins.

Structures of antibiotics mentioned in this chapter are available in appendix 1.

How Bacteria Synthesize Proteins and How Antibiotics Interfere with That Process

Bacteria need proteins, just as we do, to perform essential functions and to act as structural components of the cell. Thus, it is not surprising that the process of bacterial protein synthesis has been a major target for antibiotics. A list of some widely used protein synthesis inhibitors and their brand names is provided in Table 6.1.

Although bacteria make proteins the same way human cells do, the components of their protein-synthesizing machinery are different enough from those of our cells to make it possible to develop chemical compounds that interfere with bacterial protein synthesis but do not have the same effect on our cells. With one exception, all currently available antibiotics that inhibit bacterial protein synthesis do so by binding to ribosomes, the organelles that string amino acids together into proteins. Ribosomes are complex structures, made of 52 proteins and 3 kinds of RNA (ribosomal RNA, or rRNA). The structure of a ribosome is shown in Fig. 6.1.

Table 6.1 Widely used protein synthesis inhibitors and their brand names

Family	Antibiotic	Brand name
Aminoglycosides	Streptomycin	Streptomycin
	Amikacin	Amikin
	Neomycin	Neosporin
Tetracyclines	Doxycycline	Monodox, Vibramycin
	Oxytetracycline	Terramycin
	Demeclocycline	Declomycin
	Minocycline	Minocin
Macrolides	Erythromycin	Erythrocin
	Azithromycin	Zithromax
Lincosamides	Clindamycin	Cleocin
Streptogramin	Quinupristin + dalfopristin	Synercid
Oxazolidone	Linezolid	Zyvox
Mupirocin	Mupirocin	Bactoban

The job of a ribosome is to take an mRNA molecule and translate it into a protein (Fig. 6.2). The order in which the amino acids are strung together into a protein is critical for the functioning and stability of the protein. To ensure accurate translation of mRNA into a protein, the nucleotides in the mRNA are arranged in triplets called codons. Each codon stipulates a particular amino acid, so the order and sequence of the codons in the mRNA determine the order in which amino acids are added. The mRNA molecule snakes through the groove that is formed when the large and small subunits of the ribosome come together to form the ribosome. RNA molecules called transfer RNAs (tRNAs) come to the ribosome. Each tRNA carries a specific amino acid. The tRNA makes contact with the appropriate codon on the mRNA, and the ribosome takes the amino acid and adds it to the growing protein. The ribosome moves along the mRNA, a process called translocation, until it comes to a codon called a stop codon that signals that the end of the protein has been reached.

Before this process can occur, each amino acid has to be attached to the appropriate tRNA. This attachment step is carried out by enzymes called tRNA synthetases. These enzymes are the target of an antibiotic called mupirocin, which inhibits their activity. This antibiotic target, which is not a part of the ribosome, is the aforementioned exception to

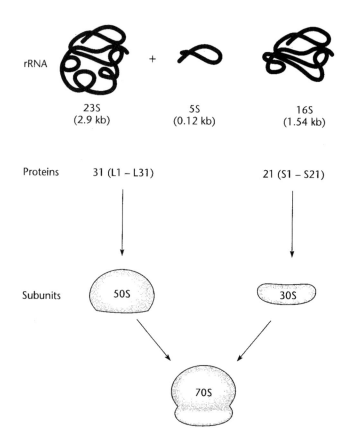

Figure 6.1   Structure of a ribosome. The bacterial ribosome is made up of two subunits, a 30S and a 50S subunit. The 30S subunit contains rRNA plus 31 ribosomal proteins. The 50S subunit contains rRNA plus 21 ribosomal proteins. The size of the complete ribosome (the 30S plus the 50S subunit) is 70S. (Reprinted from L. Snyder and W. Champness, *in Molecular Genetics of Bacteria*, ASM Press, Washington, D.C., 1997.)

the rule that protein synthesis-inhibiting antibiotics bind to the ribosome. Mupirocin was ignored for years because it is too toxic for internal use. It has been used primarily as a topical antibiotic to eliminate antibiotic-resistant *Staphylococcus aureus* from the noses of hospital workers. People colonized with *S. aureus* can transmit these bacteria to patients. In recent years, pharmaceutical companies have begun to take another look at mupirocin as a model for an important new class of antibiotics that interfere with protein synthesis.

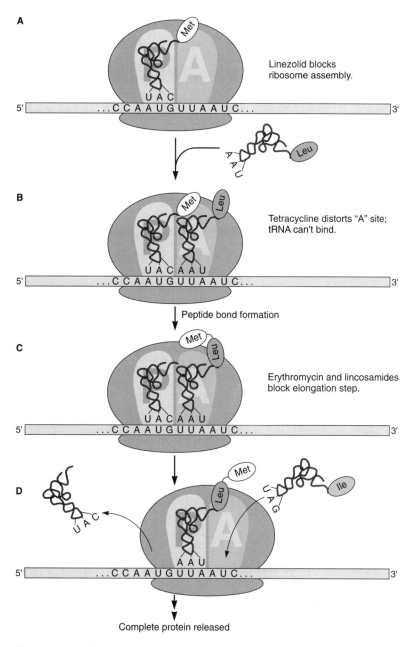

Figure 6.2   Overview of translation. After the tRNA containing methione (Met) binds to the start codon (A), the incoming tRNA bound to its amino acid (Leu) enters the A site on the 30S ribosome (B). (C) An enzyme on the 50S ribosome binds the next amino acid to the growing polypeptide. (D) The tRNA is moved to the P site, making room at the A site for another tRNA. Antibiotics that block various steps in translation are indicated.

Streptomycin and Other Aminoglycosides

Aminoglycosides are among the most widely used antibiotics. Streptomycin was one of the earliest antibiotics to enter the market, and it won its discoverer, Selman Waxman, a Nobel Prize. A special exhibit commemorating his accomplishments has been established at Rutgers University in the room that used to be his laboratory. One of the authors (A.A.S.) had the honor, as American Society for Microbiology president in 2001, of presenting a plaque designating the laboratory a microbiology historical site to the Waxman laboratory.

The first step in protein synthesis, after the attachment of the amino acids to their corresponding tRNAs, occurs when the two subunits of the ribosome assemble on one end of the mRNA molecule, at a site called the ribosome-binding site (Fig. 6.2). The antibiotic streptomycin interferes at this point. It binds to the small subunit of the ribosome and freezes the ribosome on the mRNA so that it does not proceed to synthesize the protein.

Streptomycin is a member of a family of antibiotics called aminoglycosides. This name comes from the fact that the antibiotic is composed of sugars with amino groups attached to them. Oddly enough, aminoglycosides other than streptomycin, such as amikacin, kanamycin, and neomycin, seem to prevent later steps in protein synthesis, despite the fact that they have similar structures. Nonetheless, all the members of this antibiotic family function by binding to ribosomal proteins and interfering in some way with the synthesis of bacterial proteins. Streptomycin binds to a protein in the small subunit of the ribosome. Other aminoglycosides bind to different ribosomal proteins, some in the small subunit and some in the large subunit.

One way bacteria can become resistant to aminoglycosides is to mutate the ribosomal protein that provides the binding site for the antibiotic. The antibiotic no longer binds to the ribosome and thus no longer inhibits growth of the bacteria. Such mutations are generally not seen in clinical settings because mutating ribosomal proteins makes the bacteria less able to survive in the challenging world of the human body unless the antibiotic is present. Thus, mutant strains soon disappear when antibiotic use is ended.

The fact that different aminoglycosides bind to different ribosomal proteins means that mutation of one ribosomal protein does not confer resistance to all members of the antibiotic family, yet there are strains of bacteria that are resistant to many aminoglycosides. Not surprisingly, these bacteria use another resistance strategy. Many of them have acquired an enzyme that modifies aminoglycosides by covalently attaching a chemical group (phosphoryl, acetyl, adenyl) to the antibiotic. Some examples are shown in Fig. 6.3. Modification of the antibiotic prevents it from binding to the ribosome and thus eliminates its ability to stop protein synthesis.

Bacteria can also become resistant to aminoglycosides by failing to take them up. Aminoglycosides are charged molecules that do not diffuse readily through the cytoplasmic membrane of a bacterium. Thus, they can enter the bacterial cell only if they are actively transported through the membrane. This type of resistance is not well understood because scientists still have not identified the transport proteins that transport aminoglycosides into the bacterial cytoplasm.

Aminoglycosides can have toxic side effects. In particular, they cause kidney damage and damage to the inner ear. The damage to the inner ear can result in permanent loss of hearing and loss of balance. Hearing damage can occur in as many as 10% of the patients treated with them. Fortunately, the initial stages of damage are reversible. Thus, if loss of hearing is detected, cessation of antibiotic use usually results in full recovery. Only if use continues for a prolonged period will the damage become

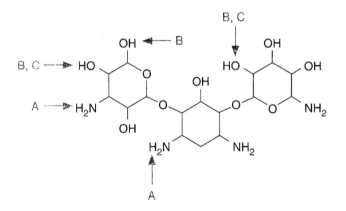

A, acetyl; B, adenyl; C, phosphoryl

Figure 6.3 Modification of aminoglycosides by attachment of chemical groups such as phosphoryl, acetyl, and adenyl. (Reprinted from A. A. Salyers and D. D. Whitt, *in Bacterial Pathogenesis: a Molecular Approach*, ASM Press, Washington, D.C., 2002.)

irreversible. Kidney damage follows the same rule. Monitoring possible side effects closely makes these antibiotics safe to use because a timely change in therapy can prevent long-term damage.

Aminoglycosides are not readily absorbed from the intestine, so they are not administered orally unless the goal is to treat bacteria in the intestinal tract itself. Most aminoglycosides are well tolerated over the short term when administered intravenously or intramuscularly. An exception is neomycin, one of the few antibiotics sold across the counter. The fact that neomycin is sold across the counter despite being too toxic to be administered internally may seem counterintuitive until one realizes that the form of neomycin being sold is an ointment for cuts and scratches. Neomycin ointment is widely available in pharmacies and is a popular therapy for preventing infections of minor wounds.

Tetracyclines

Tetracycline, like streptomycin, binds to the small subunit of the bacterial ribosome. The tetracycline family, which includes such antibiotics as doxycycline, oxytetracycline, and demeclocycline, gets its name from its structure, which consists of four fused cyclic rings. In contrast to streptomycin, tetracycline does not freeze the ribosome on the mRNA. Rather, it distorts the structure of the small ribosomal subunit, so that the incoming amino acid-bearing tRNA molecules can no longer interact properly with the mRNA. Normally, the tRNA carrying its amino acid first enters a site on the small subunit called the A site. When tetracycline binds to the small subunit of the ribosome, it distorts the A site so that the aminoacyl-tRNA cannot enter the site.

There are three known mechanisms of resistance to tetracycline. Tetracyclines diffuse readily through membranes, so bacteria cannot become resistant to tetracycline by failing to take up the antibiotic. The first mechanism of resistance to be discovered was a mechanism called antibiotic efflux. This mechanism is mediated by a protein located in the bacterial cytoplasmic membrane that actively pumps tetracycline out of the bacterial cell (efflux pump, Fig. 6.4). This mechanism of resistance has been seen as a cause of resistance to other antibiotics. It has even been implicated in resistance of cancer cells to antitumor drugs.

A second type of resistance is mediated by a bacterial protein that protects the ribosome by modifying it so that it no longer binds tetracycline but is still able to synthesize proteins (ribosome protection-type tetra-

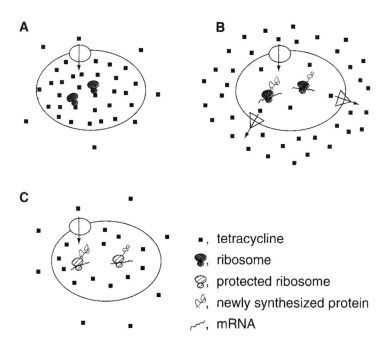

Figure 6.4   Mechanisms of tetracycline resistance. (A) Tetracycline (black squares) is taken up by a transporter (open ellipse); intracellular concentration becomes higher than extracellular concentration; tetracycline binds to ribosomes and stops protein synthesis. (B) Cytoplasmic membrane protein (open triangles) pumps tetracycline out of the cell as fast as the transporter takes it up; intracellular concentration remains too low for effective binding to ribosomes. (C) Tetracycline accumulation within the cell is similar to that in a sensitive cell, but the ribosome is protected (hatching), so tetracycline no longer binds to it. (Reprinted from A. A. Salyers and D. D. Whitt, *in* *Bacterial Pathogenesis: a Molecular Approach*, ASM Press, Washington, D.C., 2002.)

cycline resistance). Much less is known about this type of resistance mechanism than is known about the efflux pumps, but the ribosome protection type of resistance is, if anything, more widespread in bacteria than the efflux pumps. The protein that mediates the ribosome protection type of tetracycline resistance is closely related to, and probably evolved from, a protein called elongation factor G, which plays an essential role in the elongation of the protein being synthesized. This is one of many examples of how a bacterial protein that now confers resistance to an antibiotic evolved by mutation from a protein that was originally instrumental in catalyzing a part of the process targeted by the antibiotic.

A third type of resistance is of questionable clinical significance, but we mention it here because it illustrates that there are still a lot of unexplained aspects of resistance to antibiotics. This form of resistance is mediated by an enzyme that chemically modifies tetracycline, rendering it inactive. In contrast to other examples of resistance by inactivation of the antibiotic, this enzyme defies explanation.

To date, this type of resistance has been found only in strains of *Bacteroides* species, a group of bacterial species that normally live in the human intestinal tract but can also cause infections. What is strange about this enzyme is that possession of it does not make *Bacteroides* strains resistant to tetracycline. The reason is that the enzyme requires oxygen to inactivate the tetracycline molecule. *Bacteroides* species are obligate anaerobes, which cannot divide in the presence of oxygen. The gene encoding the inactivating enzyme makes *Escherichia coli* resistant to tetracycline if there is plenty of oxygen available, a situation that *E. coli* does not ordinarily encounter in the human body, where most oxygen is bound up in hemoglobin.

Where did this enzyme come from and what is its real function? No one knows. This example is interesting because it raises the very real possibility that the proteins we call resistance proteins may actually have some other function, a function we do not understand. If this is the case, there may well be a major reservoir of potential resistance proteins in nature that evolved to serve other functions but could emerge at any time in conditions involving the use of antibiotics. If our efforts to slow the emergence of resistance to antibiotics are to be successful, we need to understand such resistance anomalies.

Tetracycline has been on the market since the late 1940s and has been one of the most widely used of all antibiotics. In recent years, its popularity has declined because so many bacteria have become resistant to it. When resistance to tetracycline first began to appear, new forms of tetracycline such as doxycycline, minocycline, and oxytetracycline were effective against tetracycline-resistant bacteria. Today, many bacteria have become resistant to all of these forms of tetracycline. A new tetracycline variety that seems to thwart existing resistance mechanisms, the glycyl-glycyl tetracyclines, is now in clinical trials, so the tetracycline family may be about to stage a comeback.

A second reason for the declining popularity of tetracycline is a feature of this antibiotic that makes it unsuitable for use in infants and children. Tetracycline is a yellow compound that incorporates readily into growing bones and teeth. This incorporation has no adverse effects on the child's general health, but it does have a cosmetic effect. Tetracycline discolors teeth, giving them a permanent yellowish color. Tetracycline is a very inexpensive antibiotic and has been widely used in developing countries to treat childhood infections. Accordingly, you are most likely to see the telltale staining of teeth in adults who were children when tetracycline use in such countries was at its height.

As indicated by the example used to begin this chapter, orally administered tetracycline is still widely used by dermatologists to treat acne and another skin condition, rosacea (reddening of the skin that looks like a permanent blush). People taking tetracycline for a skin condition often take it for long periods of time, sometimes for years. This type of use may have contributed to the increase in tetracycline-resistant bacteria.

Another little noticed use of tetracycline is the use of oxytetracycline by fruit growers to prevent a disease called fire blight. The disease gets its name because the infection causes the trees to turn brown and dry, as if they had been burned. Spraying oxytetracycline on the trees prevents this disease very effectively. Another agricultural use of tetracyclines has been as growth promoters in animal husbandry. Animals fed very low doses of tetracycline tend to gain weight more rapidly. Although the increase is small, only about 4 to 5%, the margin of profit for most meat producers is so small that this makes an important economic difference.

Despite widespread bacterial resistance to tetracyclines, this group of antibiotics is still effective against some serious diseases. Two examples are anthrax and Lyme disease. As we explained in chapter 4, the postal workers who were given tetracycline to prevent inhalation anthrax probably fared better than the congressional workers who received the much more expensive antibiotic Cipro, which is no more effective against the bacterium that causes anthrax but has more side effects.

Erythromycin and Other Macrolides

Another widely used class of antibiotics is the macrolides. Erythromycin is an example of a macrolide. Macrolides have been used for years to treat a variety of bacterial infections, including respiratory infections and wound infections. The macrolides have been particularly useful as an alternative treatment for patients who are allergic to penicillin. Another use is the treatment of gastric ulcers. The macrolide clarithromyin is now being used as part of the antibiotic therapy used to cure gastric ulcers.

Erythromycin and other macrolides have had an excellent safety record, with few side effects. The main side effects are diarrhea and abdominal pain. Recently, however, a clinical study has linked macrolide use with increased risk of heart disease. What is ironic about this finding, which is only an assertion about an association, not a proof of cause and effect, is that macrolides were actually being considered for prevention of heart disease. At least some forms of heart disease appear to be caused or exacerbated by the bacterium *Chlamydia pneumoniae*, a common cause of mild respiratory disease. Erythromycin is effective against *C. pneumoniae.*

Erythromycin acts by binding to the large subunit of bacterial ribosomes and preventing the elongation of bacterial proteins. There are two other types of antibiotics, the lincosamides and the streptogramins, that do not resemble erythromycin structurally but have the same mechanism of action. They are also thwarted by the same mechanisms of resistance. Clindamycin, the antibiotic mentioned in the acne case, is a lincosamide.

Azithromycin (Zithromax), one of the newer macrolides, has been attracting attention as a treatment for sexually transmitted bacterial diseases (STDs) such as gonorrhea, chlamydia, and syphilis. People who staff STD clinics will tell you that their main concern in cases where patients have bacterial STDs is patient compliance. The clinics have the antibiotics that will cure these infections, but unless the full course is taken, the antibiotics can be ineffective. Patients who come into the clinics may have a drug or alcohol problem or may be prostitutes. Even if told to take even a week-long course of antibiotics, such patients may stop taking the antibiotic the minute symptoms subside, an outcome that occurs before the bacteria are completely eliminated. Worse, they may continue to be sexually active immediately after leaving the clinic.

In this context, Zithromax seemed like a godsend. One or two pills, administered while the patient was still at the clinic, successfully treats the patient's infection. The problem, at least for inner city clinics, is that Zithromax is a new drug still under patent control. Translation: expensive. This is the quandary faced by STD clinics, especially in low-income areas. How can you practice the best medicine for your patients and their potential sexual contacts within your budget? On the other side, how can a pharmaceutical company afford to continue to develop and test new antibiotics like Zithromax if the company cannot recoup the cost?

While the staff of inner city STD clinics is worrying about whether their budgets can support the purchase of Zithromax for patients who are likely to transmit STDs, macrolide anxiety is also occurring in a very different setting—on farms across the United States and in the offices of organizations that represent farmers. The macrolide tylosin has been widely used in swine production to prevent the bacterial respiratory diseases that afflict pigs. Even if these infections are not fatal, they retard the growth of the pigs, thus further reducing the already very slim profit margin for the farmers. Tylosin is also used, to a lesser extent, as a growth promoter.

The problem is that tylosin, although it has a different name and somewhat different structure from erythromycin, is similar enough to erythromycin to select for resistance to erythromycin and other macrolides used for humans. Tylosin also selects for resistance to lincosamides and streptogramins.

How can the same resistance mechanism work against these three different types of antibiotics? The reason is that the macrolides, lincosamides, and streptogramins, despite their different structures, bind to the same or overlapping regions on the large subunit of the bacterial ribosome. A large-subunit rRNA molecule, the 26S rRNA molecule, forms the center of the binding site of all three classes of antibiotic. Thus, a bacterial enzyme that methylates a key residue on this rRNA molecule simultaneously confers resistance to all three classes of antibiotics by reducing their binding to the ribosome.

Clindamycin: an Antibiotic That Cures but Sometimes Kills

Lincosamides are antibiotics that bind the large subunit of bacterial ribosomes. A lincosamide that has been extensively used in the therapy of human diseases caused by anaerobic bacteria is clindamycin. Not until the 1980s did clinicians accept that bacteria that could not divide in the presence of oxygen (obligate anaerobes) could cause serious human infections. It took a while to realize that the human body is not as aerobic as you might think because most of the molecular oxygen in the body is bound to proteins such as hemoglobin. Moreover, damaged tissue that loses its blood supply becomes completely anoxic very rapidly. Obligate anaerobes from the human intestine or the human mouth very quickly infect such sites if they have the opportunity.

The most common cause of this type of infection is a major genus of human intestinal bacteria, *Bacteroides*. Such infections, which can occur after abdominal trauma caused by surgery or accidents, can be deadly. Also, *Bacteroides* are naturally resistant to such antibiotics as aminoglycosides and are widely resistant to many other antibiotics, including the tetracyclines and penicillins. Clindamycin burst on the infectious disease scene as the antibiotic that would solve the problem of infections caused by anaerobic bacteria.

The problem with clindamycin and other antibiotics that are effective against obligate anaerobes is that they act not only on the bacteria that are infecting tissue and the bloodstream, but also on bacteria that normally occupy such sites as the human colon. An unwelcome side effect of clindamycin rapidly emerged. The success of clindamycin not only in treating the infection at hand but also in decimating the normal microbiota of the intestine left a window of opportunity for pathogenic bacteria that were once minority members of the intestinal microbiota to overgrow and wreak havoc. Some people, fortunately a minority, who are colonized with the pathogen *Clostridium difficile* experienced an overgrowth of these bacteria. The result was a condition called pseudomembranous colitis, a potentially fatal condition in which the colonic wall is extensively damaged by toxins produced by *C. difficile*.

Pseudomembranous colitis is not a rare curiosity. It has killed a lot of patients who were taking antibiotics that decimate the major colonic bacterial species. At one time, it seemed to be under control, but recent statistics indicate that it may be on the rise again. Does this mean people should stop taking these antibiotics? Not at all. It means that more attention needs to be paid to the fact that antibiotics are not really magic bullets after all. When a physician treats an infection, the antibiotic does not just target that infection; it may also affect the natural bacterial populations of the body that are usually protective. This realization has led to a broader view of antibiotic action and side effects.

Synercid: the First of the Human Use Streptogramins

An antibiotic called Synercid, which is a combination of two compounds, quinupristin and dalfopristin, entered the human clinical drug market very recently. Quinupristin and dalfopristin are streptogramins. They are produced by a *Streptomyces* species, *Streptomyces pristinaspiralis*. Streptogramins are not new antibiotics, but they are new to human use. A streptogramin called virginiamycin has been used for a number of years as a growth promoter in animal husbandry.

Synercid's claim to fame is that it is effective against many of the same gram-positive bacteria that were once susceptible to vancomycin. When resistance to vancomycin began to emerge, especially among *Staphylococcus* and *Streptococcus* species, the need for antibiotics that would be effective against these newly resistant gram-positive bacteria became particularly urgent. Synercid was supposed to be an answer to this problem.

In practice, Synercid has been something of a disappointment. First, Synercid is not effective against all gram-positive bacteria. In particular, *Enterococcus faecium*, a cause of postsurgical infections in immunocompromised patients, varies in its susceptibility to Synercid. Second, a side effect in some people is arthritis and general muscle pain, which can be severe. Irritation at the site of entry of the catheter also occurs in patients receiving intravenous administration via catheter. Nonetheless, Synercid could be useful in treating infections caused by vancomycin-resistant strains of *S. aureus* and *Staphylococcus epidermidis*.

Streptogramins, like the macrolides and lincosamides, act by binding to an RNA component of the large subunit of the ribosome. The fact that their binding site overlaps that of macrolides and lincosamides has an ominous implication for development of resistance to them. As already mentioned, a modification of the RNA in the large subunit of the ribosome can lead in a single step to resistance to all three classes of compounds. In fact, one type of erythromycin resistance gene, called *erm* genes, is also called MLS type resistance because the gene encodes an enzyme that modifies the ribosomal target of macrolides (M), lincosamides (L), and streptogramins (S). Thus, the useful life of Synercid could be rather short as this type of MLS expands to cover the new antibiotics.

Synercid also illustrates a frustrating problem that confronts scientists and farmers who use antibiotics to improve animal husbandry. In contrast to the animal use antibiotic tylosin, which was known to be in the same macrolide family as erythromycin and thus had the potential to select for cross-resistance to human use antibiotics, the streptogramins used in agriculture had no human use analog until Synercid entered the market. Recent studies have found animal strains of gram-positive bacteria that are resistant to Synercid, raising serious questions about how long farmers are going to be allowed to use virginiamycin. The list of agricultural antibiotics that are considered safe in the sense that they do not cross-select for resistance that could impair important human use antibiotics continues to contract.

Oxazolidones: a New Class of Protein Synthesis Inhibitor

Linezolid and eperezolid are two of the newer antibiotics to hit the market. They were synthesized in the late 1980s and are now available commercially. This class of antibiotics made news in the scientific journals because they act at an earlier stage of protein synthesis than any previously known antibiotic. Like streptomycin, they inhibit the formation of the intact ribosome (the initiation complex).

The fact that they have a different ribosomal target from previously known antibiotics is important because, in contrast to the streptogramins, these are not likely to be targets for the development of cross-resistance selected by other classes of antibiotics binding to the same site on the ribosome. Bacterial resistance to these new antibiotics will undoubtedly emerge, but we do not yet know what form the resistance mechanism will take.

The oxazolidones are used primarily to treat infections caused by gram-positive bacteria that are resistant to other antibiotics. The good news is that linezolid is proving to be more human-friendly than Synercid. Even in severely ill patients in hospital intensive-care wards, who often have numerous other underlying conditions, the incidence of severe side effects has been gratifyingly small.

The Big Picture

The antibiotics that inhibit protein synthesis illustrate better than any other types of antibiotics the complexities that are emerging in the current antibiotic use picture. First, they run the entire spectrum from the macrolides, which have been used widely to treat all sorts of infections, to newer drugs like Synercid and linezolid, which are being reserved for treatment of infections from gram-positive bacteria that are resistant to most other antibiotics. Second, the streptogramins illustrate the problem faced by farmers because the antibiotics they use are becoming moving targets. Changes in the availability of new human use antibiotics place older agricultural-use antibiotics, such as virginiamycin, in jeopardy of being banned from future agricultural use because they may cross-select for resistance to antibiotics recently added to the human use list of pharmaceuticals.

Bacterial resistance to the MLS group of antibiotics illustrates how bacteria, in a single stroke, can evolve a mechanism of resistance (in this case modification of an rRNA) that confers resistance to more than one class of antibiotic. This theme is also seen in the efflux pumps, first discovered by scientists probing mechanisms of tetracycline resistance. Now some efflux pumps are being discovered that are able to eject antibiotics of different types. We are learning that bacteria need not react to one antibiotic at a time but are capable of carrying out a multifront war against antibiotics.

Issues to Ponder

1. Given the fact that we now know that some antibiotics (e.g., macrolides and streptogramins) can cross-select for resistance to more than one class of antibiotics, should the criteria for approval of antibiotics for use in agriculture be changed, or should approval continue to be decided one antibiotic at a time? The answer to this question is not as obvious as it might seem, because cross-resistance usually does not apply to all members of all the antibiotic families involved. That is, erythromycin-resistant bacteria that have acquired an *erm* gene (MLS-type resistance) do not become resistant to all macrolides, lincosamides, and streptogramins. Similarly, the broad-spectrum efflux pumps that pump out more than one class of antibiotic are generally not universally effective. Yet.

2. There is also the more basic ethical issue of how to rank the goals of having cheap and widely available meat, which contributes to the general health and welfare of the entire population, and treating desperately ill people who already have one or more underlying problems and constitute a minority of the population. This question not only is at the base of debates over agricultural use of antibiotics but extends into the debate about how to orient research on antibiotic discovery. Should discovery efforts be targeted to achieve the greatest benefits for the human population at large or should they focus on treatment of its sickest and most compromised members? Ethically, all of us would prefer to come down on the side of doing both. Practically, we may be faced in the future with a decision between the two because of the enormous cost of developing new antibiotics.

3. To what extent do farmers have a vote in debates about antibiotics like tylosin and virginiamycin? They are a small minority of the population in terms of actual votes in an election, but they are absolutely critical to our survival. How do we weigh the need of farmers to make a decent

living against the possibility that some of their practices might, but have yet to be proven to, increase the speed with which antibiotic-resistant human pathogens emerge?

4. How are we going to rank our public priorities? Examples are the need to balance spending money to aggressively pursue the discovery and testing of new antibiotics with other claims on the public purse, such as improving education, building roads, and so on. Like it or not, such choices are going to be made and actually are already being made, but so far the public has shown very little interest in this problem.

7

Fluoroquinolones, Sulfa Drugs, and Antituberculosis Drugs

This chapter takes a look at antibiotics that have targets other than the bacterial cell wall and bacterial ribosomes. Although the antibiotics described have different targets, one thing they have in common is that resistance to them usually arises as a result of mutations in the target of the antibiotic. The balance here for the bacterium is a tricky one. The targets of antibiotics, including those described here, tend to be essential proteins. Thus, mutating them is dangerous for the bacterium, since it can result in an essential protein that is now nonfunctional.

If the mutation makes only a subtle difference, however, a difference that still allows the protein to continue to do its job but makes it less able to bind the antibiotic, the bacterium becomes resistant and still survives. Such mutations are, unfortunately, all too easily made by many bacteria. Initially, the mutation may make the bacterium somewhat less able to compete with other bacteria in the absence of the antibiotic. However, if the bacterium has the time and the antibiotic exposure to reward further mutations, compensating mutations in other proteins that interact with the protein target of the antibiotic, the originally mutated protein may come to fit better into the activities of the bacterial cell.

Structures of antibiotics mentioned in this chapter are available in appendix 1.

Fluoroquinolones: a Widely (Too Widely?) Used Family of Antibiotics

In late fall 2001, federal office workers who had worked in buildings that received anthrax-laced letters lined up to receive Cipro, the trade name of the antibiotic ciprofloxacin. Other uses of ciprofloxacin include prevention of bacterial infections in cancer chemotherapy patients, treatment of postsurgical infections, and treatment of urinary tract infections. If your physician has prescribed an antibiotic for you recently, chances are good that it was Cipro or some other fluoroquinolone.

Ciprofloxacin and other fluoroquinolones are among the most frequently prescribed antibiotics today, due in no small part to aggressive marketing by drug salesmen. These drugs are a relatively new entry into the antibiotic armamentarium and are currently regarded as one of the front-line antibiotics to defend patients against infections caused by bacteria that are resistant to other antibiotics. Also, fluoroquinolones are now being used to treat tuberculosis, a very dangerous disease, especially in cases where the bacterium that causes tuberculosis has become resistant to the traditional antituberculosis (anti-TB) drugs. Overuse of this very important type of antibiotic is already giving rise to resistance to the fluoroquinolones, an alarming development, to say the least.

The widespread use, some would say abuse, of fluoroquinolones is not limited to human medicine. In 1996, the FDA approved the use of the fluoroquinolone enrofloxacin by chicken farmers. Fluoroquinolones are administered to chickens in their water to prevent infections caused by *Escherichia coli*. Such infections can wipe out an entire flock or at least slow their growth, which is almost as bad an outcome. Thus, farmers want to treat an entire flock if even a single bird is diagnosed with this type of intestinal disease or if the disease is present in the area.

When the decision to approve enrofloxacin was made, concern about a rise in resistance to fluoroquinolones was not nearly as high as it is today, and the farmers' organizations made a persuasive case, based on scientific evidence available at the time, that the use of fluoroquinolones in agriculture was safe. Subsequently, scientists at the Centers for Disease Control and Prevention have documented a rise in fluoroquinolone resistance in *Salmonella enterica* serovar Typhimurium, a common cause of food-borne infection. This rise in resistance began at about the same time the antibiotic was introduced into agricultural use.

Of course, this increase in resistance also started about the same time that fluoroquinolones began to be used widely to treat human disease, so the contention that agricultural use of fluoroquinolones is primarily responsible for the rise in the number of resistant *Salmonella* strains is, understandably, controversial. How would using enrofloxacin to prevent an *E. coli* infection in chickens affect *S. enterica* serovar Typhimurium strains that might infect humans? *Salmonella* serovar Typhimurium, the cause of salmonellosis, also colonizes chickens, although it does not cause disease in chickens the way it does in humans. *S. enterica* serovar Typhimurium would thus be exposed to the same antibiotic used to prevent another type of infection in the chickens.

To see how complicated the arguments for and against enrofloxacin use by chicken farmers can be, consider that antibiotics are seldom used to treat patients with salmonellosis, because it is a self-limiting infection. However, in some people, a tiny minority in an outbreak, the infection moves from the intestinal tract into the bloodstream. Such a dispersed infection can be lethal. Children and the elderly are the most likely to develop the bloodstream form of the infection.

In some ways, the emphasis on fluoroquinolone resistance of *Salmonella* strains has distracted attention from a potentially more serious problem, a rise in the resistance of other bacteria such as *Enterococcus* species that are already resistant to many other antibiotics and can cause life-threatening postsurgical infections in humans. There is controversy about whether fluoroquinolone-resistant *Enterococcus* species that are found in chickens can colonize humans if the bacteria are ingested in food. If they do, they could ultimately cause serious postsurgical infections. Where do we draw the line on the agricultural use of antibiotics?

Those who want absolutely no risk to human health advocate a ban on enrofloxacin use by chicken farmers, a move that would probably result in an increase in the price of chicken for consumers and would be hard on farmers. Those who are willing to tolerate a manageable level of risk might focus on the benefits of the antibiotic use to farmers and ask what the true risk of this type of use to treatment of human infections is. Such people are in trouble, however, if they want to base their risk assessment on scientific data, because scientists are currently unable to put believable numbers on the risks associated with agricultural use of fluoroquinolones or any other antibiotic due to lack of adequate data.

It is clear that people are not currently dying by the thousands because of fluoroquinolone-resistant salmonellosis. This picture of fluoroquinolone resistance may change, however, in the future, especially if agricultural use of fluoroquinolones gives rise to resistant human pathogens that cause postsurgical infections or infections in cancer patients. Will fluoroquinolone-resistant enterococci and other human pathogens for which antibiotic treatment is an absolute necessity begin to take a greater toll? The answer to this question is a clear "yes," since fluoroquinolone-resistant pathogens already are among us and already have begun to cause medical problems. The question is whose fluoroquinolone use is to blame: that of physicians, that of farmers, or both?

Development of the Fluoroquinolones

Fluoroquinolones are frequently described as one of our newest front-line antibiotics. Technically this is true because fluoroquinolones have only recently become widely used, but a close relative, the original quinolone antibiotic, nalidixic acid, has been around since the 1960s. Although it was used successfully for treating urinary tract infections, such high doses were needed that toxic side effects were common. Its greatest use was in scientific experiments, where it was used to stop bacterial DNA replication. Often, scientists of the time wanted to know whether a process they were studying was dependent on DNA replication. Nalidixic acid was just the ticket.

Simply adding a fluorine atom to the basic quinolone structure made these antibiotics much more effective against bacteria, so that much lower concentrations could be used for therapy. Due to the lower concentration needed for efficacy, side effects became less common. Side effects do occur, however. The government workers who were given Cipro and told to take it for 60 days commonly experienced nausea and diarrhea as the course of therapy progressed. Many stopped taking the antibiotic after a few weeks.

Another side effect of fluoroquinolone use is selection for fluoroquinolone-resistant bacteria. Unfortunately, resistance to fluoroquinolones develops fairly readily, and, of course, the speed with which resistant strains appear is proportional to the amount of the antibiotic used. Because fluoroquinolones have been more aggressively marketed than most other antibiotics rather than being reserved for treatment of the most-serious human infections, fluoroquinolones may turn out to be the class of antibiotic that was the most rapidly lost—a dubious distinction indeed.

How Fluoroquinolones Act and How Bacteria Become Resistant to Them

An earlier statement that fluoroquinolones inhibit DNA replication might seem to suggest that these antibiotics inhibit the copying of a DNA strand during replication, but the mechanism of action of the fluoroquinolones is a bit more complicated than that. As a bacterium elongates prior to dividing to form two daughter cells, the bacterium's chromosome is copied by DNA polymerase so there will be a chromosome for each of the daughter cells. In order for DNA polymerase to gain access to the DNA it is going to copy, the tightly wound strands of DNA have to be unwound.

After the copying has been done, the DNA of the two resulting copies of the chromosome must be wound up again. This winding, called supercoiling, is so tight that the DNA double helices collapse into a knot-like structure. Such tight winding is necessary to make the long DNA strands compact enough to fit inside the bacterial cell. If the chromosome of a bacterium were not supercoiled, it would be many times longer than the bacterial cell.

DNA gyrase is an enzyme that is involved in returning newly replicated DNA to its supercoiled form. Fluoroquinolones act by binding tightly to DNA gyrase. In the process, they distort its structure sufficiently to prevent the gyrase from doing its job. Human cells have enzymes with similar functions, but they are different enough from the bacterial DNA gyrases that the human cell enzymes do not bind fluoroquinolones and are thus not affected by the antibiotic.

Another target of the fluoroquinolones is an enzyme called topoisomerase IV. This enzyme also plays a role in compacting bacterial DNA so that it will fit properly in the newly divided bacterial progeny. Even though there is more than one target for fluoroquinolones, mutation to resistance develops all too easily.

Bacteria can become resistant to fluoroquinolones by making one or a few mutations in one of the genes that encode a DNA gyrase subunit. The antibiotic no longer binds the mutant enzyme. Such mutations occur as frequently as one in 10 billion bacteria. This mutation rate may seem low, but bacteria are often present in the human body in concentrations that are hundreds to thousands of times higher than this.

Sulfonamides and Trimethoprim

Another commonly prescribed antibiotic combination is sulfanilamide and trimethoprim, sometimes referred to as sulfa-trimethoprim. Some trade names for this combination are Bactrim, Septra, Cofatrim, and Primsol. Sulfonamides and trimethoprim interfere with the production of the essential vitamin tetrahydrofolic acid by mimicking the substrates of two different enzymes in the pathway. Figure 7.1 shows the structures of sulfanilamide and trimethoprim along with the natural compounds they mimic. The enzymes try to carry out their normal reactions but mistakenly use the antibiotic instead of the true substrate for the reaction. The antibiotic inhibits the activity of the enzyme so that instead of smooth and efficient synthesis of tetrahydrofolate, the reaction stops.

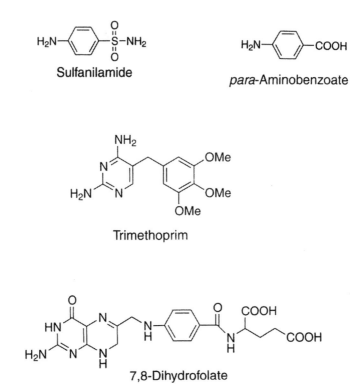

Figure 7.1   The structure of trimethoprim and sulfanilamide and the natural precursors they mimic. The natural precursors involved in the synthesis of tetrahydrofolate are *para*-aminobenzoate and 7, 8-dihydrofolate. (Reprinted from C. Walsh, *in Antibiotics: Actions, Origins, Resistance*, ASM Press, Washington, D.C., 2003.)

Tetrahydrofolic acid is a vitamin that is essential for the growth of bacteria. We humans are equally dependent on folic acid derivatives, but the sulfa-trimethoprim combination isn't toxic for us because we acquire these compounds from our diet rather than synthesizing folic acids ourselves.

As was the case of the fluoroquinolones, resistance to sulfa drugs or trimethoprim develops easily if these antibiotics are taken separately. The protein target of the antibiotic mutates so that it still works but no longer binds the antibiotic. The solution to preventing development of resistance, as well as to increasing the efficacy of the preparation, was to combine the sulfonamide with trimethoprim. The two compounds inhibit different enzymes involved in tetrahydrofolate synthesis. The probability of a bacterium becoming simultaneously resistant to both antibiotics was far lower than the probability of mutation to resistance of one of them.

A caveat is that this strategy only works if the antibiotic combination is applied simultaneously. If a patient first takes one antibiotic and then takes the second one, all bets are off, because the probability of mutation to resistance once again goes back to the dangerous one in a billion level as the enzymes mutate one at a time. In the case of Bactrim or other sulfa-trimethoprim combinations, this is not a problem because the combination is administered in a single pill.

Anti-TB Drugs

Tuberculosis (TB) is an old enemy of the human race, but it is not just a disease of the past. It is currently combining with human immunodeficiency virus (HIV) in Africa to kill millions of people, and it is once again widespread in the former Soviet Union and in Southeast Asia. In Europe and the United States, TB, which was once under control, has broken out again and is being seen more commonly despite the availability of an effective therapy. Moreover, in the developed countries, a new kind of TB threat has emerged, caused by strains of the TB bacterium (*Mycobacterium tuberculosis*) that are resistant to one or more of the anti-TB drugs. The developed countries now have a new export: drug-resistant TB strains.

Streptomycin

Streptomycin, an antibiotic that made its debut in the previous chapter as an inhibitor of bacterial protein synthesis, was the first breakthrough in effective TB therapy. Initially, the discovery of streptomycin caused tremendous excitement in the medical community. The excitement soon cooled, however, because streptomycin did not always effect a cure. Remission occurred in some patients. The problem in these patients was that *M. tuberculosis* became resistant to streptomycin during the course of therapy.

In hindsight, this was not surprising, because of the long duration of TB therapy. It was necessary to administer the antibiotic for many months, in some cases over a year, because once *M. tuberculosis* gets a foothold in the lungs, it is very difficult to eliminate. Within the lungs of an infected person, there are two populations of bacteria, one that divides rapidly and one that divides much more slowly. The acute symptoms of TB are caused by rapidly dividing *M. tuberculosis* cells. These rapidly dividing bacteria are rapidly killed by the antibiotic. The subset of the population that is growing much more slowly, however, is less susceptible to the antibiotic because lower rates of protein synthesis means lower numbers of ribosomes, which are the target of streptomycin. These bacteria persist and allow the disease to make a comeback even if the infected person's immune system is successful in controlling the initial phase of the disease. Most antibiotics are effective primarily against rapidly dividing bacteria.

The solution has been to administer a cocktail of three or more antibiotics over a period of at least 6 months. The antibiotic mixture not only is more effective than a single antibiotic against the rapidly dividing population of bacteria but also is needed to eliminate the slow growing bacteria. Also, as in the case of sulfa-trimethoprim, a mixture of antibiotics can prevent the development of resistance. Despite problems with side effects, the combination therapy has been very effective in curing patients of TB. In this case, *curing* means completely eliminating the bacteria from the lungs.

The TB-Specific Antibiotics

Some of the antibiotics included in such cocktails have familiar names, such as fluoroquinolones and streptomycin, antibiotics that are used to treat many bacterial infections. Other anti-TB antibiotics have less familiar names because they are TB specific. These antibiotics include isoniazid, ethambutol, and pyrazinamide.

These latter antibiotics are *M. tuberculosis*-specific because they inhibit production of a type of bacterial surface lipid that is peculiar to *M. tuberculosis* and only a few other types of bacteria. The lipid is called mycolic acid. It is responsible for much of the damage to the lungs that occurs when *M. tuberculosis* takes up residence. It helps cause the formation of the lesions called tubercles that destroy lung tissue and undermine lung function. Thus, inhibition of mycolic acid synthesis seriously interferes with the ability of the bacteria to cause lung damage and at the same time presumably inhibits their growth so that they can be killed by the cells of the host defense system.

Although scientists know that synthesis of mycolic acid is the target of the TB-specific drugs, it is frustrating how little is known about the details of how this occurs. Once TB was in abeyance in the developed countries (thanks to these drugs), there was little interest on the part of the funding agencies and the scientific community in doing further work to understand their mechanisms of action in more detail or to learn how *M. tuberculosis* becomes resistant to them. All of this changed when TB staged a spectacular comeback in the United States and other developed countries and resistance to the TB-specific drugs began to appear. Heroic and very expensive public health responses to the appearances of these resistant strains in New York City and other big cities have temporarily brought TB back under control, but this is probably only a temporary victory over a disease that has killed more people than all the wars ever waged and is still out there in undiminished strength—a rap sheet to be taken seriously.

Isoniazid

Isoniazid (isonicotinic acid hydrazide, INH) is an antibiotic that specifically inhibits the growth of *M. tuberculosis* and some other closely related species. It does not, for example, inhibit the growth of *Mycobacterium leprae* (the cause of leprosy) or a group of mycobacteria called atypical mycobacteria, which have become an increasing problem for immunocompromised people.

One reason for this specificity seems to be, in part, that the administered form of the antibiotic has to be activated by the bacteria before it can act. Activation occurs because the susceptible species of *Mycobacterium* have an enzyme called catalase-peroxidase that normally protects the bacteria from peroxide by converting peroxide to water, but the enzyme can also oxidize INH into the active form of the drug.

Another reason that INH acts on mycobacteria but not other bacteria is that activated INH inhibits the synthesis of the mycobacterial lipid mycolic acid, which is a major part of the mycobacterial cell wall. Mycolic acid contributes to the ability of *M. tuberculosis* to damage the infected lung.

Resistance to INH occurs and has compromised the treatment of TB because even a partial diminution in the susceptibility of the bacteria to one member of the anti-TB drug cocktail impairs the effectiveness of the total therapy. The mechanism of resistance to INH is still under study, but two mechanisms of resistance have been identified so far. One is mutations that inactivate the catalase-peroxidase enzyme, thus impairing the ability of the bacteria to activate INH. To the extent that this enzyme protects the bacteria from killing by the human phagocytic cells that defend the lungs, such a mechanism of resistance will make the bacteria somewhat less able to cause disease, but the bacteria can still do a lot of damage. A second mode of resistance is through mutations that alter one or both of some key enzymes in the mycolic acid pathway and make them less susceptible to inhibition by activated INH.

Pyrazinamide

Pyrazinamide has a structure resembling nicotinamide, a compound that participates in many bacterial processes, both catabolic and biosynthetic. Like INH, pyrazinamide is first activated by the bacteria themselves to the active form of the drug. The enzyme that carries out the activation, PZase, is probably normally involved in nucleotide metabolism, but it can also hydrolyze pyrazinamide to pyrazinoic acid, the active form of the drug. The target of pyrazinoic acid has not been identified.

Not surprisingly, a mechanism of resistance to pyrazinamide is lowered or absent PZase activity. Thus, an obvious solution would seem to be to use pyrazinoic acid rather than pyrazinamide. Unfortunately, pyrazinoic acid is not as effective as pyrazinamide, possibly because it is not taken up as well by the bacteria as pyrazinamide.

An interesting feature of pyrazinamide is the way the normal lifestyle of the bacteria makes it susceptible to pyrazinamide. *M. tuberculosis* invades the phagocytic cells of the lungs that are supposed to protect the lungs from bacteria. These cells ingest the bacteria, placing them in membrane-covered vacuoles. The interior of the vacuole then acidifies, a step that makes the killing power of the granules that subsequently fuse with the vacuole much greater. *M. tuberculosis* has developed strategies for avoiding the fate intended for them and can then break out of the vacuole and start dividing in the lung cell cytoplasm.

Unfortunately for the bacteria but fortunately for us, their uptake and accumulation of pyrazinamide increases under acidic conditions. Thus, pyrazinamide is thought to act mainly on bacteria that have invaded the phagocytic cells of the lungs. This feature makes it particularly valuable because the bacteria that manage to invade and divide inside the phagocytic cells are the ones that cause most of the lung damage and escape from other arms of the immune system by hiding inside lung cells.

Ethambutol

Until recently, little was known about the mechanism of action of ethambutol. Scientists now believe that ethambutol, like INH, interferes with the synthesis of an important component of the mycobacterial cell walls.

This component is not mycolic acid but another mycobacterium-specific cell wall compound called lipoarabinomannan, a complex lipid-polysaccharide molecule. Although ethambutol acts primarily on mycobacteria, it has a broader host range than INH and can inhibit the growth of a number of species of *Mycobacterium*.

One mechanism of resistance has been identified: mutations in the mycobacterial enzyme that reduce the binding of ethambutol to the enzyme. This is clearly not the only mechanism of resistance, however, because nearly one-third of all ethambutol-resistant strains do not have alterations in this enzyme.

Rifamycins

The most widely used member of the rifamycin family of antibiotics is rifampin (or rifampicin), but its use is restricted to only a few diseases. The most important use of rifampin is in the treatment of TB. Rifampin is also given to people in an area where an outbreak of bacterial meningitis has occurred to prevent them from contracting the disease.

Rifampin prevents bacterial growth by inhibiting the enzyme complex (RNA polymerase) that makes RNA copies of segments of the bacterial DNA genome. These RNA copies are subsequently translated to produce proteins or play other roles in protein synthesis. Resistance to rifampin arises readily from mutations in a subunit of RNA polymerase that prevents rifampin from binding to it. Such mutations occur very readily. This is one reason rifampin is used in combination with other anti-TB drugs such as INH, pyrazinamide, and ethambutol.

Other members of the rifamycin family are being evaluated for their ability to act against rifampin-resistant *M. tuberculosis*. It appears that some of them are still effective against strains that have mutated their RNA polymerase to resist the effects of rifampin.

The Rise of Resistance to Anti-TB Drugs: Connecting the DOTS

For decades, the cocktails of anti-TB drugs remained highly effective. While it is true that some strains of *M. tuberculosis* became resistant to one of the drugs in the cocktails, this phenomenon was rarely seen. Then, starting in the 1980s, resistance began to appear and accelerate. By the 1990s, drug-resistant TB had become a serious problem in the United States and other developed countries. TB has always been a major cause of infectious disease deaths worldwide but had been thought to be under control in developed countries. To have TB not only stage a comeback in developed countries but also emerge as increasingly resistant to previously effective treatments was troubling, to say the least.

What happened? Once again, complacency had raised its dangerous head. In the past, when TB was taken seriously in the United States, a public health infrastructure existed that sought and treated the disease very aggressively. Most people do not comply very well with a drug regimen that forces them to take several pills daily or 2 to 3 times a week for more than 6 months. Also, the drugs have side effects such as nausea for many people, a feature that makes compliance even poorer if patients take the drug unsupervised.

The solution was to compel TB patients to come regularly to a clinic or to have health workers visit them in their homes. The pills were given to the patient, and the health worker watched to make sure that all of them were taken. This strategy has been called DOTS (directly observed therapy short course). DOTS was highly effective, and as long as this program was in place resistance did not arise and virtually all treated people were cured of the disease.

Starting in the 1970s, however, public officials decided that since TB was under control, the relatively small number of cases did not merit the expensive infrastructure necessary to sustain a DOTS program. Dismantling of the anti-TB infrastructure began. Health workers were no longer recruited for DOTS programs, X-ray machines used to diagnose TB were given away by community health institutions, and doctors were not trained as aggressively in how to diagnose and treat TB. Even worse, work on developing new anti-TB drugs virtually ceased, as did efforts to understand how the existing anti-TB drugs worked.

All might have been well, at least for a while, if it had not been for some seemingly unrelated developments such as the rise in the prison population, the increase in the number of homeless people roaming the streets and housed in crowded shelters, and the advent of HIV. In crowded conditions, TB spreads readily, so the crowded jails and homeless shelters were a TB disaster waiting to happen. Moreover, HIV-infected people were less able than uninfected people to fight off microbial infections and had to be treated much longer for TB, a year or more compared to 6 months. Abuse of alcohol and other drugs was another factor that decreased even further the likelihood that a TB patient would comply faithfully with the long and complex therapeutic regimen.

It doesn't take a rocket scientist to predict what happened. Unsupervised infected people took one drug for a while and then another, usually only irregularly. Recall that the reason the anti-TB drugs are given as a cocktail is that experience had shown that use of a single drug often led to the development of resistance. *M. tuberculosis* became resistant first to one and then another of the anti-TB medications. A frightening example of where this was heading can be seen in the appearance in the 1990s of a strain of *M. tuberculosis* called the W strain. This strain swept through New York City and other large U.S. cities. The W strain was not only resistant to most of the anti-TB antibiotics but was as virulent as other *M. tuberculosis* strains, if not more so.

Draconian intervention measures brought outbreaks of drug-resistant strains of *M. tuberculosis* under control, but this frightening story is not over. Most people who are infected with *M. tuberculosis* (90 to 95%) do not immediately develop the full symptomatic form of the disease. Rather they harbor the bacteria in their lungs for years. In some cases, stress on a person's immune system from age, cancer chemotherapy, or any of a variety of other sources allows the bacteria to break out of their dormancy. The result is called "reactivation TB," which is just as deadly and infectious as primary TB.

Many of the people who were infected with the W strain and developed symptomatic disease died. Yet there are many more who carry that strain in their lungs. These people are ticking time bombs that could go off at any moment.

The good news is that U.S. health care officials have begun to act. DOTS is back, and research on the development of new anti-TB drugs is underway. Better late than never. The potentially bad news is that the medical community is currently showing signs of being about to repeat past errors—forgetting TB once it seems to be under control again. Oddly enough, this official amnesia is being aided by the recent focus on bioterrorism. As much, if not more, effort is now being expended on a new smallpox vaccine and a new antianthrax vaccine as on preventing and curing TB. Let's not forget that the worst bioterrorist in history, and by far the most successful, is Mother Nature. Against this terrorist, constant vigilance is the only defense. Forgetting is death.

Issues to Ponder

1. You are the editor of a prestigious scientific journal and in the aftermath of the anthrax scare you receive a manuscript that describes

the isolation and characterization of a Cipro-resistant strain of *Bacillus anthracis*, the cause of anthrax. What would you do about this manuscript?

Keep in mind that isolating a Cipro-resistant mutant is very easy and is something that virtually all bacteriologists know how to do, so this manuscript would not be revealing classified information to terrorists. Also keep in mind that such a strain, since it would have a mutant DNA gyrase, might well be less able to cause disease than the original suscepti- ble strain. Yet, the average person, including the average scientist, recoils from the possibility that such a paper might be used by terrorists. Thus, you have a PR issue to contend with, not just a decision about scientific merit. How would you define the public interest in this case? There is a strong argument in favor of publication because the findings reported in the article would empower those whose job it is to protect public health. Your journal, however, could come in for a lot of criticism if it publishes the article, because reporters who do not understand the scientific argu- ments for publication could portray it as helping terrorists.

2. Should the federal office workers who might have been exposed to spores of *B. anthracis* have been given Cipro for 60 days? Tetracycline works just as well and has fewer side effects. Also, since resistance of many bacteria to tetracycline is widespread, one could argue that long- term dosing with that antibiotic would be preferable since it would do little harm by increasing the resistance threat. Are the office workers now colonized with Cipro-resistant bacteria that could come back to bite them later in the form of a Cipro-resistant postsurgical infection? How should a public policy decision of this magnitude be made in the future? Remem- ber, such decisions have to be made within hours or, at most, a day or two. Obviously, someone who has been exposed to *B. anthracis* needs to be treated right away, and the treatment will be prolonged because no one is sure how long an effective course of antibiotics needs to be. Did the postal workers, who got tetracycline, come out of this in a better state of health than the federal office workers who got the expensive front- line drug, Cipro? So far, no one has had the nerve to ask this question experimentally.

3. Should chicken farmers be allowed to continue to use enrofloxa- cin? This is another difficult judgment question. All of us want our food to be produced cheaply and without taking up too much space. In the process of achieving these goals, we have addicted our farmers to antibiot- ics because efficient animal farming means large populations of animals

housed in very crowded conditions, conditions ideal for the spread of infectious diseases. Under what circumstances would we ask farmers to go "cold turkey" and give up antibiotics? Keep in mind that some of them would go bankrupt under such a restriction. Are these simply incompetent farmers who could make a go of antibiotic-free farming if they were better managers?

This is the European solution, but before accepting the European solution, a ban on antibiotic use in agriculture, keep in mind that in Europe farms tend to be smaller and have less concentrated animal populations. The inefficiency of these farms, compared to the factory farms of the United States, has led to a perception in Europe that the best way to fight competition from U.S. farm products is to legislate a barrier (antibiotic-free meat) that would even the playing field economically for the European farmers, who are a very powerful lobby group. To what extent should considerations of this type be taken seriously in decisions about antibiotic use?

4. We were conducting a class on bacterial lung infections for medical students. One of us asked the students, "What is the main source of TB problems in the United States today?" The unanimous answer was "immigrants." It is true that many immigrants, especially those from Africa and Asia, are infected with the bacterium that causes TB, although few have active cases of TB. We thought something was wrong with this answer, however. What was our objection and why did we try to convince the students that they needed to take a broader view of the situation? Put another way, could careless use of anti-TB drugs in the United States and Europe, which has unleashed drug-resistant TB strains on the world, be considered an act (however unintentional) of international terrorism?

8

Bacterial Promiscuity: How Bacterial Sex Contributes to Development of Resistance

The Agony of Mutation, the Ecstasy of Getting the Mutated DNA from Someone Else

Antibiotics bind to and inactivate important bacterial targets such as the enzymes that assemble the bacterial cell wall or components of ribosomes that make bacterial proteins. By mutating such a target to a form that no longer binds the antibiotic, a bacterium can become resistant to that antibiotic. However, this strategy for becoming resistant to an antibiotic is fraught with danger for the bacterium. Changing such essential components runs the risk to a bacterium of rendering itself dead or severely crippled.

What is the solution, then, from the bacterial perspective? That's easy. Go to the bacterial equivalent of eBay and obtain a resistance gene that has been created by some other bacterium, probably created the hard way, by throwing the genetic dice multiple times and leaving a lot of dead bacteria along the way. Is there really a bacterial eBay equivalent? Unfortunately there is, in the form of gene swapping by bacteria, a process known to bacteriologists as bacterial sex. (Even nerds can have a little fun with scientific nomenclature!)

Bacterial Promiscuity in the Form of Rampant Gene Sharing

Even though bacteria that mutate to resistance can mutate further to become more comfortable with their newly resistant state, the process still requires multiple rounds of random mutations that can be highly deleterious to the bacterium involved, with only a few bacteria emerging with the winning set of mutations. Also, taking the random mutation route means a prolonged period of time for a bacterium to become resistant to antibiotics. Unfortunately for us humans, there is a gentler, kinder way for bacteria to become resistant to an antibiotic. This alternative strategy is to acquire a preformed gene from another bacterium. The newly acquired DNA segment seldom poses any threat to the bacterium that acquires it. Although initially it may exact a fitness cost, it is unlikely to kill the bacterium. About the worst that can happen is that the gene is not expressed at first and thus confers no benefit at all. A bacterium can easily fix this problem by acquiring mutations in the promoter region of the gene, the region that controls whether a messenger RNA (mRNA) copy is made and translated into the resistance protein. Such mutations in a foreign gene are highly unlikely to affect the fitness of a bacterium since they affect the expression of a potentially beneficial gene.

Bacteria can acquire new DNA from each other by any one of three processes: conjugation (direct cell-to-cell transfer of DNA), transformation (uptake of DNA from the environment), and bacteriophage transduction (transfer by bacterial viruses). Conjugation is the process that seems to be responsible for most of the movement of resistance genes among bacteria in nature, especially among bacteria from different species and genera.

Conjugation has been called "bacterial sex" by microbiologists. Two bacteria come together and make tight contact with each other. A multiprotein channel forms between the cytoplasms of the two cells and DNA is passed through this channel from the donor cell to the recipient cell. A copy is retained by the donor cell (Fig. 8.1). Sounds like sex to us. When conjugation was first discovered, the cell that donated the DNA was called F+ for "fertility-proficient" and the recipient cell was called F−. F+ donors were described as "males" and F− recipients as "females." To make the metaphor even more convincing, F+ cells of some strains display a long tube-shaped projection on their cell surfaces. It was immediately dubbed a "sex pilus."

Unfortunately, there are some holes in this sex analogy. First, the sex pilus turned out to behave more like an arm than a penis. It attaches to the recipient cell and retracts, thus bringing the two cells into close contact. DNA does not pass through the sex pilus but rather through a separate channel that forms between the two mating cells. Second, many bacteria capable of exchanging DNA by conjugation do not produce a sex pilus. A physicist, whose name has unfortunately been lost to science history, warned that the analogy was getting out of hand. He calculated that the movement of a long DNA molecule through a tube the size of the sex pilus would generate so much heat that the pilus would be destroyed.

In a way, it is too bad that the sex analogy seems a bit stretched,

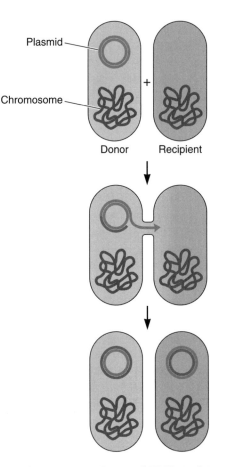

Figure 8.1   Bacterial conjugation. Bacterial DNA, in this case in the form of a plasmid, is transferred from one cell to another by cell-to-cell contact.

because the imagination could go completely wild. Bacteria can transfer DNA from one genus to another, the rough equivalent of a human having a sexual encounter with a slug. In fact, some bacteria can transfer DNA to plant cells, yeasts, and other eukaryotic cells. Now that's promiscuity! You are probably wondering whether the process of conjugation is limited to pairs of bacteria or pairs of bacterial and eukaryotic cells. Well, at least in the laboratory, three can tango.

To understand this phenomenon, it is first necessary to take a short detour into the biology of the transferred DNA segments. One type of transferred DNA element is a plasmid, a segment of DNA that can be either circular or linear and replicates autonomously. Plasmids can be thought of as mini-chromosomes, although this description drives some language purists wild because plasmids generally do not contain such housekeeping genes as the genes needed for transcription of mRNA or genes needed to produce the components of a ribosome. Plasmids capable of transferring themselves by conjugation carry the genes needed to construct the sex pilus and the multiprotein complex through which the DNA copy of the plasmid passes (self-transmissible plasmids). Plasmids that do not carry such genes can nonetheless be mobilized by self-transmissible plasmids because they can use the DNA transfer channel constructed by the self-transmissible plasmid to transfer themselves to the recipient (mobilizable plasmids).

Now, back to the promised ménage à trois, what microbiologists call triparental matings. There are two donor cells, one of which carries a self-transmissible plasmid that transmits itself to the second donor, which carries a mobilizable plasmid. After entering the strain containing a mobilizable plasmid, the self-transmissible plasmid makes the former recipient into a donor, and the self-transmissible plasmid mobilizes the mobilizable plasmid into the third strain, which is the ultimate recipient. To our knowledge, no one has tried to see whether a conjugation foursome can be demonstrated.

Resistance Gene Transfer and Agricultural Use of Antibiotics

As described in chapter 4, tons of antibiotics are used in animal husbandry each year. A substantial amount of those antibiotics (no one can agree exactly how much) is used to enhance the growth of food animals, especially pigs and chickens. Antibiotics do not necessarily increase the final weight of the animal, but rather speed weight gains by about 4 to 5%. Although the difference in the number of days needed for an antibiotic-fed animal to reach its market weight may be small, that difference is critical to farmers, who have very narrow profit margins. Another, somewhat less controversial, use of antibiotics in animal husbandry is for prevention of disease (prophylaxis). That is, there may be no clinical signs of disease or even signs that disease is imminent, but antibiotics are given to ensure that the animals remain disease free. Again, the margin of safety this gives farmers is critical in these days of tight meat markets, to ensure that farmers end up on the right side of the profit-loss boundary.

Farmers in Europe have supposedly given up the use of antibiotics to promote growth and have restricted prophylactic use. This experiment seems to have been successful, although there were some transition pains in the form of lower meat yields due to disease when the ban was first enforced. The example of the European farmers' turn away from agricultural use of antibiotics has been hailed by some as a shining example for farmers in the United States and other countries that are still holding on to the use of antibiotics as growth promoters.

There is a key difference, however, between the way animal husbandry is practiced in the United States and Europe. In the United States, consumers have demanded cheap meat prices. They also want to have animal production facilities confined to the smallest space possible, leaving more land available for houses and parks. The result has been an industry that runs farms on which animals live in crowded conditions, conditions that favor the spread of infectious diseases. By comparison, European farms house far fewer animals at lower densities.

What does this little excursion into animal husbandry have to do with bacterial promiscuity? The answer lies in the series of steps that might link the use of large quantities of antibiotics in agriculture with increasing resistance to antibiotics by bacteria that cause human disease. The scenario goes as follows. Antibiotic-resistant bacteria are selected in the intestinal tracts of animals fed antibiotics. These resistant bacteria contaminate the carcass during slaughter and enter the food supply on the meat. Evidence that this occurs is accumulating. In theory, cooking should eliminate the contaminating bacteria, but it is likely that many of them survive due to imperfect consumer hygiene and enter the consumer's colon. In the colon, the incoming bacteria encounter the bacteria that normally inhabit the human colon. Since incoming bacteria are adapted to nonhuman animals and have furthermore just undergone trial by acid as they passed through the stomach, it is likely that most of them will be unable to compete effectively enough with the resident bacteria to stay in the colon. Rather, most of them will be excreted without ceremony.

If the incoming bacteria pass right through the colon and are excreted, how could they possibly cause us any problems? The answer is that during the 24 to 48 h they spend passing through the human colon, they could transfer their antibiotic resistance genes to native human colonic bacteria. After all, transfer of DNA by conjugation takes at most a few hours. Moreover, concentrations of the native bacteria are high and many of them are located on the surfaces of plant particles or the mucin that lubricates the intestinal lining. This is the sort of close proximity that facilitates conjugation.

Although colonic bacteria are innocuous as long as they stay in the colon, they can cause deadly infections if they are released from the colon during surgery or other abdominal trauma. Having a highly antibiotic-resistant normal bacterial population makes a person a ticking time bomb, primed to go off if that person ever has the misfortune to contract a post-surgical infection, most of which are caused by members of the patient's own microbiota or the microbiota of health care workers.

How likely is it that bacteria living in the human colon would actually acquire resistance genes from bacteria that are only passing through? Asked in a different way, how much gene transfer occurs naturally among human colonic bacteria or between them and bacteria that do not permanently inhabit the site? Initially, scientists assumed that the probability of such transfers was low because in the laboratory the frequency of resistance gene transfer by naturally occurring gene transfer elements is usually quite low. For example, if an antibiotic-resistant colon bacterium (the gene donor) is mixed with a susceptible bacterium (the recipient) under optimized conditions, less than one in 10,000 of the bacteria that are potential recipients will receive the transferred gene. As the old saying goes, though, truth can be stranger than fiction.

A pair of undergraduate students in A. Salyers' laboratory (Heather Vlamakis, who has since gone on to graduate school, and Aikiesha Shelby, who has since entered medical school) aided research specialist Nadja Shoemaker in a project to find out exactly how much dalliance was actually going on in the human colon. To make the exercise easier, they chose to focus on *Bacteroides* species, a type of gram-negative bacterium that accounts for about one-fourth of the bacteria in the human colon. They looked at two sets of strains. One set of strains was isolated before 1970. A second set of strains had been isolated in the late 1990s. The research team found, to their amazement, that whereas only about 20% of the pre-1970s strains carried a tetracycline resistance gene called *tetQ*, over 80% of the strains isolated in the 1990s carried *tetQ*. The *tetQ* genes in all of these strains, old and new, had virtually identical DNA sequences, whereas the DNA sequences of the genomes of the different *Bacteroides* species involved in this study differed by a much greater margin. This finding of DNA sequence identity between *tetQ* genes in very diverse species of bacteria supported the hypothesis that the *tetQ* gene had been transferred from one strain to another, not evolved independently. That is, bacterial sex seems to have been responsible for the spread of *tetQ*.

That hypothesis was strengthened by the finding that *tetQ* was indeed carried on a conjugally transferred integrated DNA element, called a conjugative transposon. Scientists had known for some time that plasmids could transfer themselves and the genes they carry from one bacterial species to another, but the finding that conjugative transposons, a newly recognized type of conjugally transmissible element, were so active in transferring resistance genes among *Bacteroides* species was something of a surprise.

Aside from the finding that there has been a lot of conjugating going on in the colon over the past three decades, there were two other surprising findings from this study. First, the fact that 80% of *Bacteroides* strains were carrying the *tetQ* gene, even though many of the strains had been isolated from healthy people with no recent use of antibiotics, indicated that carriage of this resistance gene was very stable. This finding flew in the face of a favorite assumption on the part of physicians and many scientists that if you just stop using an antibiotic, the resistant strains disappear because they are not as fit as the susceptible strains in the absence of the antibiotic. Obviously, *tetQ* and the transmissible element that carried it were not taking enough of a fitness toll on these *Bacteroides* strains. As scientist Anne Summers once said, antibiotic resistance genes are easy for bacteria to acquire but difficult for them to lose.

A second unexpected finding was that already in the pre-1970 period, when tetracycline had only begun to be used in clinical medicine, such a high percentage of strains (20% or more) already carried *tetQ*. Where did this resistance gene come from, and what selection pressures led to it already being widespread before tetracycline use became intense? This was not the first observation suggesting that resistance genes were around before antibiotics were used widely in medicine. The implication is obvious. There may be selection pressures other than the antibiotics themselves that promote carriage of what we now call antibiotic resistance genes.

The standard explanation for this phenomenon is that many soil bacteria produce antibics such as tetracycline and that contact with these antibiotics in the soil has caused bacteria to become resistant to them even before they were harnessed by human scientists. However, this theory is not very convincing because the concentrations of antibiotics in soil from this activity are so low as to be undetectable, certainly not high enough to explain such a high level of resistance gene evolution seen in natural isolates. A more likely-sounding explanation is that there are some plant compounds such as alkaloids that resemble tetracycline or other antibiotics. These compounds would be present in significantly high concentrations in soil. The only problem with this theory is that there is not one shred of evidence to support it. Also, it does not explain resistance to completely synthetic antibiotics. Professor Julian Davies, a scientist in Vancouver, Canada, who has worked for decades on antibiotics and antibiotic resistance, has yet another theory. His idea is that antibiotics were originally signaling molecules that allowed bacteria to communicate with each other. In this view of antibiotics, the genes to which the antibiotics bound, which ultimately became resistance genes, were the receptors for these signals. There is no direct evidence for this theory except that in a few cases, antibiotics have been found to serve as inducers of gene expression in bacteria.

An interesting twist to this story is that the conjugative transposons that have been moving *tetQ* around in that swinging bacterial single's bar we know as the human colon are stimulated to transfer DNA by none other than the antibiotic tetracycline itself. Put in a different way, if you agree to call conjugative transfer of resistance genes "bacterial sex," you would have to think of tetracycline as an aphrodisiac for bacteria that carry this particular type of resistance transfer element. Antibiotics not only select for survival of antibiotic-resistant bacteria; they may also, in some cases, stimulate the transfer of the resistance gene in the first place.

Thus, in the case of *tetQ*, widespread use of tetracycline by physicians (and farmers?) could have been responsible for stimulating the transfer of *tetQ* among *Bacteroides* species. As described in chapter 6, tetracycline is not used only to cure acute infections such as gonorrhea and certain types of pneumonia; it is also used by dermatologists to treat acne and another skin condition called rosacea. Dermatology patients take tetracycline orally for months and sometimes years. In such patients, the selection pressure would thus be present for a long time.

The study just described supports the contention that horizontal transfer of antibiotic resistance genes is a lot more frequent in the colon than anyone expected, but it does not address the question raised earlier about gene transfer between swallowed bacteria or other members of the colonic bacterial population besides *Bacteroides* species. Scientists have

Table 8.1   Antibiotic resistance genes in *Bacteroides* from uninfected
people (community isolates) and infected people (clinical isolates)[a]

Source of isolates	No.	Percentage of isolates carrying			
		tetQ	*ermF*	*ermG*	*cfxA*
Community (pre-1970)	69	32	0	0	0
Clinical (pre-1970)	23	22	9	0	4
Community (1996–1997)	102	81	15	8	3
Clinical (1980–1995)	87	86	30	18	14

[a]The number of isolates tested is in the first column. The other columns contain the
percentage of that number that carried *tetQ* or other resistance genes. Note that
the incidence of strains carrying the resistance genes rose between the pre-1970s
period and the 1990s. All of the resistance genes shown here are carried on elements that can be transferred from one bacterium to another by conjugation, and
this is probably the way the genes spread.

used the same tactic as was used to detect possible examples of horizontal
gene transfer in *Bacteroides* species to look for other examples of horizontal
gene transfer between different genera of bacteria in the colon. That is,
they have looked for identical genes in very distantly related species as
an indication that some type of genetic conduit is open between the species
in which the gene is found. A summary of some of the results of this
effort is shown in Table 8.1.

The team in the Salyers' laboratory that followed *tetQ* transfer among
colonic *Bacteroides* had also noticed that a gene that conferred resistance
to the antibiotic erythromycin (*ermB*) had appeared in *Bacteroides* species
only in the post-1970s period. This gene had previously been seen only
in gram-positive bacteria such as *Staphylococcus*, *Streptococcus*, and *Bacillus*
species. Sure enough, the genes now being found in *Bacteroides* species
were more than 99% identical at the DNA sequence level to the genes that
had been found previously in gram-positive bacteria. This result strongly
supports the hypothesis that these genes were being transferred between
different genera. More to the point, since *Streptococcus* and *Staphylococcus*
species do not colonize the colon but are swallowed and are merely passing through, finding identical genes in these genera and in *Bacteroides*

species indicates that there is genetic communication between bacteria that normally reside in the human colon and bacteria that are simply passing through. Subsequently, *ermB* has been found to be carried on conjugative transposons, so they could well have been transferred by conjugation.

A New Way to Spend Your Vacation: Cruise a Pig Manure "Lagoon"

Even more recently, *ermB* and another *erm* gene, *ermG*, have been found in gram-positive bacteria isolated from a pig manure storage facility in Peoria, Ill., by U.S. Department of Agriculture scientists Terry Whitehead and Mike Cotta. Whitehead and Cotta, working together with University of Illinois graduate students Yanping Wang and Anamika Gupta, found that these genes shared over 99% DNA sequence identity with genes found in human colonic *Bacteroides* species. It seems clear, then, that at least some resistance genes are moving quite widely among bacterial species found in natural settings, including species normally found in very different environments.

It is important to note what this type of evidence does NOT show—the direction of transfer. It is not possible at this point to conclude whether the *erm* genes in pig manure moved into human bacterial strains or vice versa. It may be possible to do this in the future by using DNA sequence signatures (the less than 1% sequence variation between genes in bacteria isolated from different sites), coupled with comparisons of old isolates and new isolates, where such isolates are available, to determine the direction of transfer. Currently, the best that scientists can do is to say that there is some sort of genetic conduit between these different species of bacteria, found in different locations, that allows the movement of antibiotic resistance genes between them. To many people, this is sufficient cause for alarm.

As was mentioned earlier, the transfer of the type of conjugative transposon that carries *tetQ* and some other types of antibiotic resistance genes between *Bacteroides* species is stimulated by exposing the bacteria to tetracycline. By contrast, the conjugative transposons that carry the *ermB* and *ermG* genes seem to be able to transfer these genes without stimulation by an antibiotic. Their transfer may be stimulated by other conditions, but as yet such conditions have not been identified.

What practical use are we to make of the type of information that has been described in this and the previous sections? First, we learn that in some cases antibiotics such as tetracycline can stimulate bacteria to transfer antibiotic resistance genes. Thus, when a new antibiotic is being evaluated for possible use, more than its predilection for selecting for bacteria that have become resistant needs to be considered. The effect of the antibiotic on stimulating transfer of resistance genes also needs to be evaluated, insofar as we know enough about various resistance gene transfer elements to do that. Second, it is no longer safe to assume that, once evolved, an antibiotic resistance gene will stay within a single species and in a single location. This observation underscores the importance of limiting the use of antibiotics as much as possible to cases in which they are essential for maintaining human health.

Use of Antibiotics in Human Medicine and Resistance Gene Transfer

In June 2002 a frightening milestone was passed. By that time the antibiotic vancomycin had become the last defense against strains of *Staphylococcus* and *Streptococcus* species, species that can cause overwhelming fatal infections and that had become resistant to most antibiotics. For several years after vancomycin began to be used heavily in human medicine, these species remained reassuringly susceptible to vancomycin. The vancomycin barrier began to crack when a type of gram-positive bacteria related to staphylococci and streptococci, *Enterococcus* species, began to become vancomycin resistant.

Enterococcus, as the name suggests, is a genus of bacteria found in the colons of nearly everyone. It does not normally cause infection, but if it escapes from the colon, it can cause serious wound and bloodstream infections. Some hospitals have had very bad experiences with postsurgical *Enterococcus* infections in their intensive care wards. Even in this case, however, it was somewhat reassuring, in a backhanded way, that the *Enterococcus* species, *E. faecium*, that seemed to be the species least able to cause infection, was the one that was most likely to become resistant to vancomycin.

In a controversial laboratory experiment, a scientist had shown that *E. faecium* could transfer its vancomycin resistance genes to *Staphylcoccus aureus*. The experiment was controversial because a laboratory scientist had created just the type of *S. aureus* strain everyone feared. The strain was hastily destroyed by the investigator, but the point had been made that *S. aureus* could well become resistant to vancomycin by acquiring vancomycin resistance genes by conjugation. Recall from chapter 5 that some vancomycin resistance genes have been shown to be carried on transmissible elements such as plasmids and conjugative transposons.

Nonetheless, when transfer of vancomycin resistance genes to *S. aureus* did not happen immediately, physicians began to hope that *S. aureus* would remain obediently susceptible to vancomycin and would retain its virginity despite repeated advances from the enterococci. In early fall, 2003, this hope was shattered. The first case of a fully resistant strain of *S. aureus* was reported. The strain was isolated from a patient receiving kidney dialysis. Such patients often develop infections from mixtures of bacterial pathogens. In this case, both *S. aureus* and *E. faecium* were isolated from the patient's bloodstream. Some of the *S. aureus* isolates were susceptible to vancomycin, whereas others were resistant to vancomycin. Further investigation provided evidence that the *E. faecium* may have transferred its resistance genes to *S. aureus*, possibly within that same patient.

This is an ominous development. It is not clear whether this will prove to be an isolated occurrence or whether vancomycin-resistant *S. aureus* strains will become more and more common. Sometimes, it takes years for resistant strains to spread to many different cities and countries, but from now on, everyone is on staph alert. There is a tiny silver lining in this very dark cloud. So far, the multiply antibiotic-resistant strains of *S. aureus* and other pathogens have been susceptible to some antibiotics. Bacteria that are not susceptible to any antibiotic have been seen but are still rare. The caution here is that even if there is still one (or even three) antibiotic that works, much more laboratory work will need to be done to find that antibiotic. Meanwhile, the patient may sustain irreversible damage to important organs such as the lungs, kidney, and brain from the progression of septic shock. Some patients may die because the shock process has gone too far for an antibiotic to be effective, even if an antibiotic that is effective in the laboratory is identified. People should take little comfort in the fact that the vancomycin-resistant bacteria of today may be susceptible to some other antibiotic. There is always tomorrow.

Whether on the farm or in the clinic, bacterial sharing of resistance is a problem. The continuing rise in the number of resistant strains underscores the need for enhanced use of hygienic procedures such as handwashing and proper use of drugs to prevent the progeny of bacteria that have become resistant from spreading.

Also, it may become necessary to screen incoming patients for carriage of bacteria with certain types of resistance patterns, e.g., multidrug-resistant *S. aureus* strains, to know which patients may need to be isolated and watched with greatest care. This may be especially important for large urban hospitals with crowded intensive care wards. Such a screening program has been tried in Western Australia on an experimental basis. A team led by infectious disease specialist Peter Collignon showed that careful screening of incoming patients and follow up procedures that either cleared patients of carriage of the antibiotic-resistant bacteria or prevented them from spreading their resistant strains to other patients did reduce significantly the incidence of infections from multidrug-resistant *S. aureus*. Yes, it was a very expensive program. One hopes that this type of draconian intervention will not be necessary in the future, but don't be certain of that.

A final story about vancomycin resistance that was introduced in chapter 5 brings us back to agricultural use of antibiotics. In Europe, although not in the United States, an antibiotic called avoparcin was approved in the 1990s in the European Union for use as a growth promoter. Have you ever heard of avoparcin? Neither had most people involved in the monitoring of resistance patterns in humans. Avoparcin is an analog of vancomycin, and it selects for resistance to vancomycin as well as to itself. Not surprisingly, after a couple of years of avoparcin use in Europe, vancomycin-resistant strains of *E. faecium* began to appear in farm animals. More troubling was the fact that vancomycin-resistant *E. faecium* strains began to be found in the intestines of urban European adults.

Such resistant enterococci were not found in people in the United States, where avoparcin had not been approved for use in agriculture. Extensive vancomycin use in hospitals had spawned vancomycin-resistant *E. faecium* strains, but these seemed to be limited to hospitals and did not move into the community. Europeans, by contrast, had not had much of a problem with vancomycin-resistant enterococci in their hospitals because vancomycin use was much more severely limited in human medicine than was the case in the United States. The different ecologies of the vancomycin-resistant strains in Europe and the United States supported the hypothesis that the European urban adults who had acquired the vancomycin-resistant strains had acquired them through the food supply from farms that used avoparcin as a growth promoter. Consistent with this hypothesis, when the European Union abruptly banned avoparcin use in agriculture after being alerted to its dangers, the carriage of vancomycin-resistant enterococci in the colons of urban adults decreased immediately from about 10 to 4%. This experience, which has been followed carefully by European scientists, is one of the best examples of a possible linkage between farm use of antibiotics and antibiotic resistance patterns in human intestinal bacteria.

The Flap over Genetically Modified Foods

So far in this chapter, conjugation has been the star of the DNA transfer show, but transfer of antibiotic resistance genes can also occur by transformation, the uptake of DNA from the environment. This well-known fact led opponents of genetically modified (GM) foods to raise antibiotic resistance as a possible safety concern associated with GM foods. GM plants usually carry an antibiotic resistance gene, most often either an ampicillin resistance gene or an aminoglycoside resistance gene, because these genes have been employed as selectable markers on plasmids used to clone genes that are destined to be moved into plant cells, genes such as the gene that encodes the insecticidal toxin BT produced by *Bacillus thuringiensis*. When the plasmid carrying the BT toxin gene was moved into the plant cell, the antibiotic resistance gene went along for the ride and was integrated into the plant genome.

The antibiotic resistance gene that enters the plant genome has no effect on the plant. Not only is it a bacterial gene that is not expressed in the plant, but plant cells, being eukaryotic cells, are naturally resistant to antibiotics such as ampicillin. Plants, of course, do not have a cell wall similar to that of bacteria and are thus impervious to peptidoglycan synthesis inhibitors.

The concern expressed by the antibiotechnology activists was that the bacterial antibiotic resistance genes lodged in the plant genome might be released in an intact form from the plant cells during digestion in the human intestinal tract and be taken up by human intestinal bacteria. If the resistance genes were retained by the transformed intestinal bacteria, these bacteria might become resistant to ampicillin. Numerous groups of scientists met all over the world to discuss this possibility. Regulatory agencies in Europe, the United States, and many other countries consulted antibiotic resistance experts to assess the possible magnitude of this danger.

Almost unanimously, the microbiologists who were consulted concluded that there was little or no danger of such a gene transfer event happening. For one thing, only a small number of bacteria are capable of spontaneously taking up DNA from the environment. This is why laboratory scientists who want to introduce DNA into *Escherichia coli* have to electrocute these bacteria or otherwise force them to artificially take up exogenous DNA.

If the DNA is taken up, it will have to integrate into the chromosome in order to be retained. It can only do that through homologous recombination, a process that requires that portions of the incoming DNA are identical in DNA sequence to regions of the recipient bacterium's genome. The effect of this is that most gene transfers that occur in nature by transformation occur between very closely related strains of the same species. That is, an antibiotic resistance gene derived from *E. coli* or related species, as most of the marker genes on cloning vectors were, could not enter and survive in bacteria that are not part of this very narrow group.

Despite compelling arguments that the movement of an ampicillin resistance gene from a plant cell to an intestinal bacterium was extremely unlikely, being as it was the result of a series of highly unlikely steps, the bottom line was that scientists could not say that the probability was zero. Accordingly, after all the tortuous arguments about the frequency of the various steps from the initial DNA uptake event to the fixing of the intact gene in the chromosome of the recipient bacterium, the argument finally came down to the question of what the consequences would be if such an event occurred.

The answer to this was easy. Acquisition by an intestinal bacterium of an ampicillin resistance gene from the plasmids currently used in cloning would be the biggest clinical nonevent of the 20th (and 21st) century. The ampicillin gene and other resistance genes in cloning vectors were cloned in the 1970s. Since then, scientists have learned to deal with genes of this vintage. In fact, if you take a β-lactam preparation chances are that the preparation also contains a second compound that is an inhibitor of the type of β-lactamase that is encoded by the resistance gene on the cloning vector.

The problem today is that the genes encoding resistance to ampicillin and other members of the β-lactam family have mutated to become resistant to more advanced antibiotics than ampicillin. Some of these genes even encode β-lactamases that are resistant to β-lactamase inhibitors. Physicians would love to get the old ampicillin resistance gene such as the one found on cloning vectors back, because these are the ones physicians know how to control. A physician once told one of us that if it were ever shown that today's β-lactam resistance genes could be replaced in human intestinal bacteria by the genes found on cloning vectors, he would counsel his patients to eat as much GM food as they could stand, as a preventive measure to protect them against the much more dangerous genes that have evolved since the 1970s.

The debate over antibiotic resistance genes in GM crops did reveal one safety hazard associated with these crops. Extended frivolous debates about what was clearly not a safety problem distracted the public and, more importantly, government agencies from real antibiotic resistance problems, such as the rise of bacteria resistant to many antibiotics. It is ironic that the Europeans, the ones most obsessed with the resistance genes in GM crops, apparently were so distracted by this debate that they approved, virtually without discussion, the use of avoparcin as a growth promoter in food animals. The debate over the safety of GM plants is an instructive example of how arguments that are really more about trade restrictions and philosophical issues can distort public discourse about health issues in a way that can actually imperil human health.

Lessons from the Gene Transfer Front

Aside from learning that conjugative transfer of resistance genes occurs much more readily in nature than was previously thought, some other lessons have been learned in recent years by people working in this area. A major lesson comes from the observation that more than one type of resistance gene can reside on the same gene transfer element. To appreciate what this means, consider a gene transfer element that carries a tetracycline resistance gene and an erythromycin resistance gene. The tetracycline resistance gene does not confer resistance to erythromycin and vice versa, but if these two genes are carried on the same transmissible element, selection for tetracycline resistance becomes selection for erythromycin resistance as well because the genes are linked. Thus, contrary to the common assumption that one type of antibiotic selects only for resistance to that type of antibiotic, genetic linkage makes it possible for one type of antibiotic to select for resistance to other types of antibiotics as well.

A type of element called an integron provides an extreme, and very troubling, example of linkage. Integrons are receptacles that pile up multiple genes, all under the control of the integron-provided promoter (Fig. 8.2). Thus, an integron can carry several antibiotic resistance genes. Integrons are a form of transposon that can enter conjugative elements such as plasmids, thus allowing the multigene cassette to move to other bacteria. Perhaps the most frightening examples of integrons carrying multiple resistance genes are those that carry not only antibiotic resistance genes but also genes for resistance to antiseptics and disinfectants. If antibiotics

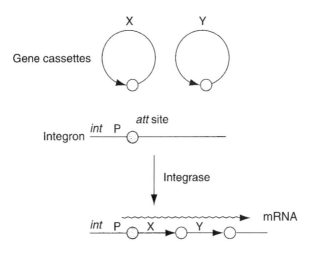

Figure 8.2   Integration of two gene cassettes, carrying promoterless resistance genes X and Y into an integron. The integron supplies the promoter (P) and an integrase gene (*int*). The shaded circle represents the insertion site. Arrows represent the direction of transcription. (Reprinted from A. A. Salyers and D. D. Whitt, *in Bacterial Pathogenesis: a Molecular Approach*, ASM Press, Washington, D. C., 2002.)

lose their efficacy because of bacterial resistance, antiseptics and disinfectants are the last remaining line of protection against bacterial infections. Usually, disinfectants such as bleach and alcohol are so general in their action that it is difficult if not impossible for bacteria to become resistant to them. However, bacteria have managed to become resistant to some types of disinfectants such as quaternary ammonium compounds. A combination of resistance to antibiotics and resistance to some antiseptics or disinfectants could be deadly, not only for hospital patients but also for sick people in the community.

Issues to Ponder

1. Now that it has become clear that bacterial sharing of resistance genes is a widespread bacterial activity, the consequences of this have to be considered. A physician or farmer who treats a particular bacterial infection also treats the microbiota of the intestine, skin, vaginal tract, and upper respiratory tract. How seriously should this fact be considered in the approval of new uses of antibiotics? Keep in mind that more and more often, the new antibiotics are desperately needed to combat infections

caused by resistant bacteria. How much time and money should be expended on evaluating the possibility that these genes might be transferred in natural settings such as the human body or environments exposed to antibiotic runoff?

2. An interesting wrinkle to the previous question became apparent in connection with probiotics, strains of bacteria that are ingested intentionally to bolster the protective effect of the intestinal microbiota. Most of these strains are members of the genus *Lactobacillus*. It turns out that most strains of *Lactobacillus* species are resistant to vancomycin. One of us (A.S.) pointed out to scientists who isolated and marketed these strains that perhaps they should find out whether this resistance was transmissible to other bacteria, such as *S. aureus* or *Enterococcus* species. Was A.S. hailed initially as a possible savior of the field from lawyers or as a villain of previously unimaginable magnitude? You guessed it. Nonetheless, in the intervening time, scientists in the probiotics field have met this challenge by showing that resistance to vancomycin in *Lactobacillus* species was a general metabolic trait connected to the cell wall structure of this genus and was thus not likely to be transmissible. Forewarned is forearmed, and there are now some companies that might have been the target of frivolous lawsuits that are now in a good position to fend off such attacks. It's not easy being a resistance crusader. Here's the question. Should A.S. have asked the probiotics scientists about vancomycin resistance, thus causing them pain and suffering that turned out to be unnecessary? Should she be sued for their pain and suffering?

3. Possessing a microbiota that contains antibiotic-resistant bacteria is potentially a problem for you if you have an operation and get a postsurgical infection caused by members of your microbiota. It is also a problem for your children. Scientists think, with some degree of certainty, that infants acquire their bacterial microbiota from their parents or other caregivers. What is the legacy we are passing on to our children? Also, should antibiotic resistance be considered a food safety issue?

9

The Looming Crisis in Antibiotic Availability

There are three obvious responses to the antibiotic resistance problem. One is to try to educate physicians and patients about the appropriate use of antibiotics, in the hope that such education programs will change the behavior of both groups. Such efforts are now being made and seem to be making a difference for the better. Nonetheless, this type of effort alone is unlikely to bring about the sort of major changes that may be necessary to preserve antibiotics.

A second response would be to limit a physician's freedom to prescribe antibiotics by placing veto power in the hands of a pharmacist or infectious disease specialist. This approach is not unprecedented. After all, there are limitations on a physician's rights to prescribe such drugs as morphine. Such an approach might work in hospitals, and some form of oversight over prescriptions has already been introduced in a few large hospitals. For example, a physician deemed to be misprescribing or over-prescribing antibiotics may receive an admonitory letter from the hospital infection control agent. The infection control agent also watches for signs of bad hygienic practices, which manifest themselves in the form of increases in hospital-acquired infections. Preventing such infections not only is good for the patient but also prevents overuse of antibiotics. This type of arrangement may work in a large hospital but would likely turn into a bureaucratic nightmare if extended to physicians practicing in the community.

A third response would be to step up the discovery of new antibiotics. Many people will be surprised to hear that this obvious technological fix is not being aggressively pursued. If anything, the movement has been in the direction of reducing research and development of new antibiotics. More and more pharmaceutical companies are shutting down or cutting back on their antibiotic discovery programs. Among the large pharmaceutical companies that have cut back on or eliminated their in-house programs for antibiotic discovery are Eli Lilly, Bristol-Meyer Squibb, GlaxoSmithKline, Aventis, Proctor and Gamble, Roche, and Wyeth-Ayerst.

Some critics of the pharmaceutical industry have attributed this development to the greed and mean-spiritedness of the big pharmaceutical companies who would rather peddle Prozac and Viagra than focus on antibiotics that save lives, but in fact the decision of the pharmaceutical companies has been based on compelling economic pressures that need to be understood if they are to be countered.

The Profitability Calculation: an Inescapable Feature of Modern Markets

It is true that the profitability of antibiotics is not as great as that of drugs for treating neurological diseases, heart disease, cancer, and depression. An indication of this is provided by a number pharmaceutical companies use to estimate the likely profitability of a drug. This number is called the net present value (NPV) and is a measure of the likely profitability of a drug. As in many other areas of life, large is good. Whereas musculoskeletal drugs, such as medications for rheumatoid arthritis, and neurological drugs, such as antidepressants, have NPVs ($\times$ \$1,000,000) of 1,150 and 720, respectively, an injectable antibiotic only has an NPV of 100. Even a setback such as the recent Vioxx debacle is unlikely to reduce this gap significantly. Pharmaceutical researcher Steven Projan (Wyeth-Ayerst) has pointed out recently that proposed increases in FDA requirements that would increase the number of people in clinical trials of antibiotics could reduce the number from 100 to 35.

Clearly, the relatively unattractive profitability of antibiotics has played a major role in the decision of pharmaceutical companies to pull out of the antibiotic discovery field. A critical factor in the cost of bringing an antibiotic to market is the cost of clinical trials necessary to demonstrate safety and efficacy. Everyone agrees that such testing is necessary, but the bottom line is that it now costs over \$800 million to take a new antibiotic through the testing and approval process (Fig. 9.1). To make matters worse, new antibiotics are becoming more and more difficult to find. A surprising and discouraging development has been that highly hyped new discovery methods, such as mining the genome sequences of bacteria or combinatorial chemistry methods for producing large collections of

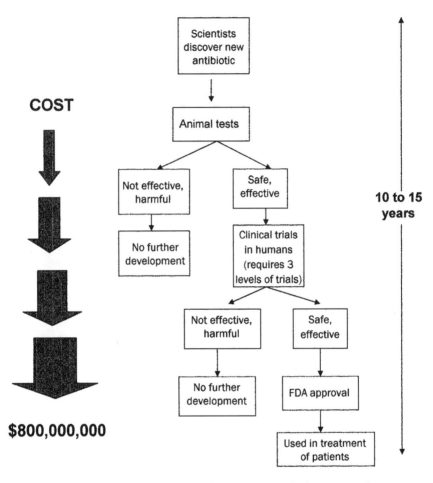

Figure 9.1   The process required for the approval of a new antibiotic.

variations on a chemical core compound, have not produced the flood of new antibiotics that was expected. It is not clear why this is, but it is a troubling development because it means that some glitzy new method is not likely to provide a quick fix. Twenty or 30 years down the line, perhaps, but not tomorrow.

The pharmaceutical companies are currently facing another type of attack. Consumers are convinced, understandably, that drug prices are too high and are trying various tactics, such as buying drugs in other countries, to reduce these costs. An objective evaluation of the contribution of drug prices to overall health costs is overdue, but figures available so far suggest that drugs are not the main contributor to the incredible recent increases in health costs. Putting aside the cost of such drugs as cholesterol-lowering medications or blood pressure medications, antibiotics certainly qualify as a major bargain. Given the benefit of a quick cure for a condition such as a urinary tract infection or ulcers, most people would not consider a few hundred dollars too great a price to pay.

Hidden behind these calculations, however, is a question that still has to be confronted by the public and the medical community. How are drug discovery and development to be financed? Right now, people in developed countries, especially the United States, are paying these costs. Should these costs be spread out over a larger base and not just folded into the cost of a drug that is still under patent protection? Could the costs of developing new antibiotics be reduced without endangering public health? As a nation and a world, we have yet to make a serious attempt to confront these critical questions. A point worth making is that the pharmaceutical companies have a disingenuous argument of their own. To hear them tell it, they bear all the costs of research and development. In fact, in virtually all cases the initial discoveries are made in academic or government laboratories or in small biotech startup companies. The big pharmaceutical companies acquire these discoveries and develop them. Also, a large part of what is called "development" by some drug companies is not just the clinical trials, but advertising the drug.

Do We Really Need New Antibiotics?

Given the expense of developing a new antibiotic, some in the medical community have questioned whether we really need any more new antibiotics. This startling view is based on the fact that at present, the vast majority of patients who need to be treated with antibiotics have an infection that can be controlled with some antibiotic. The outcome of their treatment will be satisfactory, at least for the present.

The counterargument looks to the future rather than to the present. Vancomycin was hailed as the solution to methicillin-resistant *Staphylococcus aureus* (MRSA) infections, but vancomycin-resistant strains of *S. aureus* have already begun to emerge. These strains have so far been susceptible to at least one other antibiotic, but the fact that *S. aureus* can become resistant to vancomycin raises the specter of the future evolution of a strain that is resistant to all available antibiotics. Similarly, the fact that

Streptococcus pneumoniae remained susceptible to penicillin for so long was used to argue that there was no need to consider using other antibiotics for therapy. The recent dramatic rise of penicillin-resistant strains of *S. pneumoniae* and the concomitant rise of strains resistant to macrolides and tetracycline point out the need to expect bacteria to become resistant to a trusted antibiotic at some point in the future.

There are already some strains of bacteria that have been called "pan-resistant" because they are resistant to all commonly used antibiotics. Some examples are *Pseudomonas aeruginosa*, a cause of postsurgical infections and a common cause of lung infections in cystic fibrosis patients, and *Acinetobacter baumanii*, a not-well-publicized species that is making the rounds in intensive care wards. In some cases, physicians have felt constrained to resort to antibiotics that were previously considered too toxic for general use, an ominous portent. Such cases are still uncommon, but they are occurring.

Perhaps a useful way to look at this problem is to ask what would have happened if back in the 1960s the medical community had decided that the first generation of penicillins and tetracyclines was adequate and stopped discovering new antibiotics. If that decision had been made, we would be in serious trouble today. True, some bacteria are still susceptible to good old penicillin and tetracycline, but resistance to these antibiotics has become so pervasive that we would now be facing a treatment crisis if the decision to stop with the first-generation antibiotics had been made.

Keep in mind that infectious diseases are the second leading cause of death worldwide. In the developed world, infectious diseases have been demoted to number three as the leading cause of death, but that is not much of a demotion, especially given all the advances we have made in prevention of these diseases. Infectious diseases could come roaring back to number two or even number one in the future if care is not taken now. The old science fiction movies used to end with an admonition to "watch the sky." Today, in the antibiotic resistance environment we are experiencing, the admonition is "watch the hospitals."

Given that bacteria have surprised us in the past with their adaptability and given the strong selection pressures imposed by the widespread use of antibiotics, not only in human medicine but also in agriculture, it is unlikely that bacteria will stop evolving new resistance strategies. After all, bacteria have been here for over 3 billion years, whereas we humans number our history in the millions of years. It is most unwise to underestimate the capabilities of such ancient organisms.

Another reason to look ahead is that time as well as cost is a problem in bringing a new antibiotic to market. Because the clinical trials and the subsequent approval process take time, it can take over 10 years for an antibiotic to go from the discovery phase to the market. This is not simple red tape; but is the amount of time needed for a rigorous assessment of the efficacy and safety of antibiotics that covers different age groups and people with different genetic backgrounds. Becoming complacent about the need for new antibiotics is a dangerous strategy. It assumes that disease-causing bacteria will remain pretty much the way they are now for the foreseeable future. If you believe that, you will believe anything.

Attitudes in the Medical Community

There is more to the antibiotic discovery problem than profitability. There is also the problem of the blasé attitude of the medical community toward bacterial infections. If the majority of physicians do not take bacterial diseases seriously, why should anyone else do so? The complacent attitude of many physicians had its origin, ironically, in the initial elation over the success of antibiotics and the fact that humans seemed to have gained, for the first time in history, the upper hand over our microscopic enemies.

During the 1960s, physicians and patients alike had every reason to be complacent about the continued efficacy of antibiotics. This complacency was reflected in the pronouncement that has been attributed to William Stewart, a former U.S. Surgeon General. In 1967, Stewart supposedly proclaimed that research scientists could safely forget bacterial infections and turn their attention to more pressing matters like cancer and heart disease. Those who have attempted to find any concrete evidence for this quote have come up empty-handed. Even though it is unlikely that Stewart himself actually made this pronouncement, there is general agreement that it is an accurate portrayal of the attitude of the period.

Back to Antibiotics, Again

There has been an ironic twist in recent years to such early optimism about the impending success of the war on cancer and other chronic diseases. Although there has been substantial progress in the treatment and prevention of cancer and other diseases, a general dissatisfaction has arisen about the fact that diseases like Alzheimer's, inflammatory bowel disease, and heart disease seem to be resisting the efforts of scientists to eliminate them. This has caused some groups such as the Colitis and Crohn's Disease Foundation and the American Heart Association to turn back to the bacteriologists in the hope that some forms of chronic diseases might actually be caused by bacteria. If so, antibiotics might be able to cure or prevent these diseases.

What sparked this retro-revolution was the discovery that ulcers, which had long been considered to be caused by overproduction of stomach acid, proved to be caused in most cases by bacteria. This discovery led to the development of an antibiotic regimen for curing ulcers. Thus, instead of spending thousands of dollars a year and still being plagued by ulcers, patients can now be cured of ulcers by taking a 1- to 2-week-long course of antibiotics that costs about $200.

It remains to be seen whether other diseases previously considered not to be infectious will be found to have a bacterial cause. If such discoveries are not made, it will not be because of any lack of enthusiasm and optimism among the scientists and policy makers, who have taken ulcers as a clarion call to action. Of course, there is a dark side to this story. *Helicobacter pylori*, the bacterium that causes most ulcers, is becoming resistant to the antibiotics now being used to treat ulcers. If other previously intractable diseases are added to the list of diseases discovered to be caused by bacteria, the use of antibiotics will become even more widespread and the consequent pressure for the development of resistance by bacteria will become even greater. Also, there will be more opportunities for public fury if a cure is lost.

The Hospital Administrators Weigh In

Although individual physicians have been slow to admit that increasing antibiotic resistance among disease-causing bacteria is becoming a serious problem, hospital administrators have begun to note the increased cost of treating a patient with an antibiotic-resistant bacterial infection. These increased costs are due not only to the need to use more expensive antibiotics, but also to the increased number of days spent in the hospital and later treatment of conditions arising from failure to bring an infection under control in a timely manner, conditions arising from damage to the heart, lung, or kidneys. Moreover, a hospital that has a higher than average incidence of difficult-to-treat postsurgical infections might acquire a reputation that would frighten patients away. Not incidentally, the disruption of family life for patients who suffer infections that keep them hospitalized longer than expected is a serious social concern, one that is seldom factored into the cost calculations done by the accountants.

The rising concern of hospital administrators over the cost consequences of antibiotic-resistant bacterial infections has had one good consequence for patients. It has given infectious disease control specialists increased status and power in the hospital. The infectious disease control specialists, now doctors rather than (as in the past) low-level technical support staff, have initiated new efforts to reinforce the use of simple but effective hygienic measures such as hand washing. These measures had tended to be forgotten in an era when antibiotics could be counted upon to clean up any little messes created by lax hygienic practices. Infectious disease specialists have even begun, in some settings, to isolate patients with antibiotic-resistant bacterial infections. Once this practice was limited to patients with highly contagious diseases such as tuberculosis or plague, but antibiotic-resistant bacteria carried by a person with a postsurgical infection can be spread to other patients in a crowded hospital environment, with potentially disastrous consequences.

The Lawyers Weigh In

Lawyers have also stepped into the resistance picture, for once on the side of the angels. Reportedly, there have already been a number of lawsuits against hospitals on behalf of patients who acquired bacterial infections in the hospital, especially those that proved to be resistant to antibiotics. So far, there have been few news reports of lawsuits in the United States, probably because the majority of these cases are settled out of court.

A case did make the news in Scotland in 2003. It is worth describing because it illustrates the horrific problems that postsurgical infections can cause. An article appeared in the April 8, 2003, issue of *The Scotsman* entitled "Superbug victims sue 'dirty' hospital." The article described a lawsuit being brought in Glasgow on behalf of patients and their families who have suffered from infections caused by antibiotic-resistant bacteria. One of the cases described was that of a 77-year-old woman who entered the hospital with a "relatively minor" kidney infection, then developed septicemia (bacteria in the bloodstream) and was hovering near death at the time the article was written. If she survived, it was possible that she might have to have her legs amputated because of painful sores she had developed on her feet as a consequence of the spreading infection.

This patient's infection was caused by MRSA, which is now resistant to many antibiotics in addition to methicillin (a relative of penicillin). MRSA has become the scourge of large hospitals in many parts of the world. The threat of lawsuits, sadly, may prove to be the spur that makes physicians finally start to take antibiotic-resistant bacteria seriously.

In the same article in *The Scotsman*, an infectious disease microbiologist was quoted as saying, "There is no doubt that MRSA is a marker for poor hygiene. When you have a hospital with poor hygiene, which is understaffed and overcrowded, it is a recipe for MRSA." The reporter who authored this article seemed to be confused about whether *S. aureus* was a virus or a bacterium, but the message came through loud and clear: hospitals need to be given the resources and encouragement to clean up their act, literally—or else.

Beyond Hospitals

Are antibiotic-resistant bacteria mainly a problem in hospitals and in nursing homes, where similar populations of crowded, highly susceptible people are found? Unfortunately not. If anything, the use of antibiotics in the community, outside of hospitals, is actually heavier in the aggregate than the use of antibiotics in hospitals. MRSA was at one time considered by the experts to be a hospital-specific phenomenon. Then reports of community cases of such infections began to trickle in. Scientists from Australia were among the first to report seeing patients with MRSA infections coming into hospitals from the community, and this has led some hospitals to begin screening patients for carriage of resistant bacteria such as MRSA.

A wake-up call in the United States was a report in *Morbidity and Mortality Weekly Reports* (*MMWR*), a journal of the U.S. Centers for Disease Control and Prevention (CDC). This report described four cases of juvenile deaths caused by MRSA in the U.S. Midwest from 1997 to 1999. These children had none of the usual risk factors for MRSA infections, such as a long hospital stay, a compromised immune system, or surgical wounds. At present, the community MRSA strains are more likely to be susceptible to other antibiotics than the hospital strains, but how long will this be the case? It is becoming clear that bacteria resistant to multiple antibiotics are a community problem, not just a hospital problem. This chilling realization has practical consequences. The battle to contain MRSA cannot be waged solely in the easily controlled hospital environment, but needs to be taken to the community as a whole.

At present, the main defense against the hospital variety of MRSA strains and the multidrug resistant strains of *S. pneumoniae*, the main cause of bacterial pneumonia, is the antibiotic vancomycin. The first strains of MRSA that are less susceptible to vancomycin began to appear in 2001. The first report of a fully vancomycin-resistant strain of *S. aureus* appeared in late 2003. Thus, we have to modify an earlier statement. The admonition du jour is not just "watch the skies" or "watch the hospitals"; it is now "watch your neighbor"—not a prescription for amicable social interactions.

The reader has no doubt detected the rising hysteria in the tone of the preceding text. Accordingly, a sobering and reassuring note needs to be sounded here. In most cases, as already mentioned, even the increasingly vancomycin-resistant strains of MRSA have proved to be susceptible to some other antibiotic. This is not an invitation to complacency, however, but only a time-out during which decisive action can still be taken. If such decisive action is not taken, we face the possibility of the emergence of the first fully antibiotic-resistant strain, resistant to all available antibiotics, of a very serious bacterial pathogen. Does this mean back to the mercury ointments?

The Tricky Question of How Safe Is "Safe"

The question of whether antibiotic testing procedures could be streamlined in some way was raised earlier. In fact, the trend lately has been to make the safety and efficacy testing for antibiotics more stringent. The FDA requires pharmaceutical companies to test enough patients in the final, large clinical trials to reach a degree of certainty that an antibiotic is safe and effective. The standards set by the FDA in the past have been a model for state-of-the-art safety standards. Is there, however, a point of vanishing benefit to increasing these standards further?

Recently, this issue surfaced in the form of an FDA proposal to decrease even further the margin of error in determining whether an antibiotic is safe and effective. The bar was already high, but the proposal was to raise the bar even higher. The effect of this increase in the number of subjects tested would be to increase the cost of clinical trials substantially, making antibiotics even less attractive to the pharmaceutical companies. As mentioned earlier in the chapter, this change could reduce the NPV of an antibiotic from 100, already a dangerously low figure, to 35.

This is a hard issue for the public to decide. If asked whether you would like your drugs to be even safer and more effective, who among us would say "no"? But if you were asked whether a small increased margin in safety and efficacy is worth driving more pharmaceutical companies out of the antibiotic discovery area, the answer would probably be different. The problem here for the public is that the two contenders both have a vested interest in the outcome. Some independent source of information is badly needed to help the public decide who is right, the pharmaceutical companies who have given us so many life-saving and life-enhancing drugs or the FDA, which has an enviable history of protecting the public.

Another bone of contention between the pharmaceutical companies and the FDA is the FDA's decision to require that a new antibiotic must be demonstrated to be clearly superior to existing antibiotics. This sounds reasonable enough until one realizes that an antibiotic that is equally effective or even slightly less effective than those currently on the market for treating a certain type of infectious disease can suddenly become clearly superior if the bacteria involved become resistant to the antibiotics already on the market.

Vancomycin serves as a good example of this. Initially, vancomycin was considered to be less effective than existing antibiotics because it targeted gram-positive bacteria only, yet when gram-positive bacteria became a serious resistance problem, vancomycin's attractiveness improved significantly. Clearly, the existence of the resistance problem makes antibiotics materially different from products that treat conditions such as heart disease or depression, where sudden development of resistance is not a problem.

What Is To Be Done?

If the flow of new antibiotics from the research laboratory to the market is being impeded by economic pressures on for-profit companies, who will take up the slack? Currently, the word on the antibiotic street is that the government, in the form of the National Institutes of Health (NIH) and the CDC, may begin to play a bigger role in the antibiotic discovery effort. In fact, the NIH, in a reorganization of its infectious diseases research funding (the Institute for Allergy and Infectious Disease) has created a new panel to consider grant proposals on antibiotic discovery and resistance.

How can the pharmaceutical companies be drawn back into the antibiotic discovery area? One thing that needs to be done is to stop demonizing the industry as profit-hungry predators. Although we would be the last people to christen Big Pharma as the Mother Teresa of industries, the industry has largely behaved in a responsible way and has certainly contributed to improving the quality of human life. And, after all, the pharmaceutical companies are for-profit companies with shareholders who are very interested in stock prices and profitability.

One suggestion for encouraging the pharmaceutical companies to re-enter the antibiotic discovery area is to make antibiotics "orphan drugs." Orphan drugs are ones that are important but have too small a market to be profitable. This would be hard to justify, however, because of the huge volume of antibiotic sales. A more adventurous suggestion has been to extend patent protection for companies that develop a new antibiotic. This suggestion, which has been called "wild card exclusivity" would not necessarily apply to the antibiotic itself but would allow a company to extend patent protection for 6 months or so on the drug of its choice, perhaps one of the profitable rheumatoid arthritis or anti-depression drugs.

Patients who are seeing their medical costs spiraling out of control might find this "pill" hard to swallow because it would keep drug prices higher longer, but drugs are a relatively minor component of medical costs as a whole. Whatever you think about this type of proposal, it is important to keep in mind that solving the antibiotic discovery problem is not going to be simple and it is not going to be cheap. What the public is willing to accept in the end comes down to how valuable the average person thinks antibiotics are.

Where Have All the Experts Gone?

There is another problem that will not be solved, at least in the short term, by throwing more money at antibiotic development efforts. In recent years, the study of bacteria has become somewhat unfashionable in academic circles, and the number of scientists being trained in this area has declined to alarmingly low levels. This is ironic because it was the bacteriologists who not only discovered antibiotics but, more recently, created the molecular revolution that has transformed all of biology. In fact, bacteriologists have a track record that is unequaled in biology.

There is, however, a silver lining in this otherwise dark cloud. The molecular revolution has spawned a renewed interest in how bacteria cause disease, a field called bacterial pathogenesis. In fact, we just happen to be the authors of another book called *Bacterial Pathogenesis*, which every reader of this book should now rush out to buy. Seriously, however, due to the new interest in bacterial pathogenesis, more students are now being attracted to this area, so the population of bacteriologists may soon once again reach critical mass. It will take some time to replenish the ragged and aging band of experts that currently exists, but there is reason to think that such replenishment is beginning to occur.

Issues to Ponder

1. How do you feel about the apportionment of resources in medical research? The resources being brought to bear by developed countries are impressive but are finite. It may prove to be the case that increased investment in the discovery and development of new antibiotics, whether by the government or by the pharmaceutical industry, reduces the amount of money spent on stem cell research or other promising, but not proven to be successful, future therapies for diabetes and Alzheimer's. To what extent do you think funding for research on antibiotic discovery and antibiotic resistance should be favored over other types of research? Assume that developed countries like the United States are currently devoting about as much money to medical research as the economists tell us is feasible now and in the future. Thus, resources added to one area detract from resources given to other areas.

2. A sore subject among microbiologists interested in human health is the shocking amount of money that is currently being diverted into bioterrorism research. This is definitely having a negative impact on the funding of research on much more common and more problematic diseases such as bacterial pneumonia, MRSA, and, on the viral side, influenza and HIV. The anthrax attacks were a problem largely because public ignorance of the nature of the problem, augmented by the silence of the U.S. government, caused a panic that was not justified given the small number of people infected. In the past, we have seen the same funding pattern with Hanta virus (anyone remember that?). How should we change the way decisions are made about the distribution of federal funding for research so that a more prudent course is taken and political concerns do not disrupt important research initiatives?

3. We have tried to present all sides of the issues involved in the

decisions made by pharmaceutical companies about where to invest research and development funds and decisions made by regulatory agencies such as the FDA about how to regulate antibiotics. A couple of questions arise from the material presented here. First, given that pharmaceutical companies are for-profit organizations, how (assuming you think this is a good idea) can they be coaxed back into the antibiotic discovery business? Extended patent protection? Orphan drug program? Other options? Alternatively, should the National Institutes of Health, which is currently doing little about antibiotic discovery, be directed to increase its attention to this issue? Again, keep in mind that resources are finite, so realistically speaking, resources directed toward antibiotic discovery would reduce resources currently flowing into other areas.

4. The need for new antibiotics is not a U.S. problem. It is an international problem. What contribution should other countries be making? After all, some pharmaceutical companies are owned by multinational companies. Should the United Nations become involved?

10

Antiseptics and Disinfectants

Role in Human Health

The normal defenses of our bodies are remarkably effective in keeping bacteria that impinge on us daily from causing disease. But they are not 100% effective. Not only have bacteria developed ways of circumventing our bodies' defenses, but intentional disruption of these defenses, such as surgery or cancer chemotherapy, lays the body temporarily open to infection. One of the greatest advances in human health during the past century was the discovery that our natural defenses could be augmented with externally provided chemical defenses: antiseptics, disinfectants, and antibiotics. This book has focused on antibiotics because of their great importance in curing and preventing infection, but it is now time to look at two other very important types of antibacterial compounds: antiseptics and disinfectants. These antimicrobial compounds may well prove to be our best response to bacteria that are resistant to antibiotics.

Structures of antiseptics and disinfectants mentioned in this chapter are available in appendix 1.

Mechanisms of Action

Antiseptics and disinfectants, like antibiotics, are chemicals that kill or inhibit the growth of bacteria and other microorganisms. Most antiseptics and disinfectants are bactericidal. Most are also effective against other types of disease-causing microbes such as viruses, fungi, and protozoa. This broad coverage has a drawback, however, because the chemicals used as antiseptics and disinfectants are too toxic for internal use in humans. Accordingly, they are applied only to skin or inanimate surfaces. *Antiseptic* is the term used to describe antimicrobial compounds applied to skin, e.g., in handwashing preparations used in hospitals and doctors' offices. *Disinfectant* is the term used to describe antimicrobial compounds applied to inanimate objects and surfaces. Some compounds, such as iodine and quaternary ammonium compounds, fall into both categories. Others, such as household bleach (chlorine) and some forms of phenol, are too harsh for use on skin and are used only as disinfectants. Similarly, alkylating agents such as formaldehyde are too toxic for application to skin.

Antiseptics and disinfectants tend to attack multiple targets in microbes. For example, iodine and chlorine are strong oxidants that inactivate many microbial proteins. Hydrogen peroxide is also a strong oxidant that inactivates many microbial targets. In addition to inactivating proteins, halides and peroxide can also damage microbial DNA, so they have multiple killing functions. These compounds do not just act on microbial proteins and DNA. They can damage human cells too. This is why when they are used on humans they are applied to skin, the surface of which consists of dead cells and is thus not affected by the toxic activities of such antiseptics.

Antiseptics can also target membranes. For example, quaternary ammonium compounds (QACs), one of the most widely used types of antiseptic, insert themselves into phospholipid bilayer membranes, causing cells to leak vital ions and other small molecules. These compounds also disrupt electron transport chains that provide energy for the metabolic activities of microbial cells. Table 10.1 summarizes the mechanisms of action of some common antiseptics and disinfectants.

Table 10.1 Modes of action of common antiseptics and disinfectants

Antiseptic/ disinfectant	Mode of killing	Examples of products
Alcohols	Denature proteins, destroy membranes	Ethanol added to mouth-wash, isopropanol
Alkylating agents	Inactivate proteins	Formaldehyde, ethylene oxide (gas used for sterilizing inanimate objects)
Halides (I^-, Cl^-)	Oxidize proteins, damage DNA	Household bleach (Cl), iodine for cuts
Heavy metals (Hg^{2+}, Ag^+)	Denature proteins	Silver in catheter plastic, Thimerosol
Phenols	Denature proteins, disrupt cell membranes	Hexachlorophene (Phisohex)
Quaternary ammonium compounds	Disrupt cell membranes	Zephiran, benzalkonium bromide, Cetrimide

Antiseptics and disinfectants do best against actively replicating microorganisms. Thus, bacterial spores, inert survival forms that some bacteria can assume to weather adverse environmental consequences, are generally resistant to them. Bacterial biofilms are also more resistant than free-living bacteria. Antiseptics and disinfectants are effective against a wide range of fungi, protozoa, viruses, and bacteria. Their lack of specificity for a particular type of microbe makes them useful as all-purpose microbe killers.

In almost all cases, antiseptics and disinfectants are benevolent agents that, when properly used, make an enormous contribution to protecting people, especially those facing surgery, from contracting an infection. In one case, however, an antiseptic has proven to have unexpected adverse side effects. Phisohex, a phenolic antiseptic, has long been widely used for cleansing the skin of teenagers battling acne or patients preparing for surgery. Up to 20 years ago, Phisohex was also used widely in hospitals to bathe newborn babies. What was not appreciated at the time was that babies' skin is more absorbent than adults' skin. This fact, together with the immature state of the covering of the brain and spinal cord immediately following birth, apparently led to some neurological complications in some infants. Although it was not established conclusively that Phisohex was the culprit, use of Phisohex for washing infants and any children below the age of 2 years has been discontinued. There is no evidence of any adverse effects on adults. Today, Phisohex is sold only by prescription.

Abuse of Antiseptics and Disinfectants

Some people can't get enough of a good thing. Just as physicians and their patients overuse and abuse antibiotics, consumers have shown a propensity for overusing antiseptics and disinfectants. Things have gotten to the point that it is difficult to find a hand soap that does not have "antibacterial" on the label. The "antibacterial" label has also popped up on cutting boards and children's toys. In hospitals, antibacterial hand cleansers are very important for protecting the health of patients and health care workers alike, but there is little or no evidence that these same products have any useful role in the average household. For one thing, it is questionable whether antibacterial household products have any significant protective effect. An antibacterial compound incorporated into a cutting board is certainly not going to remove all the *Salmonella* from a cutting board that has just been given a cursory wipe between cutting chicken on it and cutting vegetables for a salad. For another, most people do not lather up and rub their hands for over 15 seconds with their "antibacterial" hand soap, as hospital workers and physicians are trained to do.

There are two good reasons for NOT using antibacterial products. First, prolonged exposure of bacteria to antibacterial compounds is likely to select for bacteria that are resistant to antiseptics and disinfectants. This is definitely not a good outcome, given that antiseptics and disinfectants may soon be our last defense against some antibiotic-resistant strains of bacteria. Second, scientists are beginning to suspect that keeping children too clean may be bad for them. The human body evolved over millions of years to withstand a heavy load of bacteria. Additionally, some components of the immune system, which was designed mainly as a protection against microbes, can misfire, causing allergies and asthma. The suspicion is strong that such misfirings are more likely to occur if the immune system does not have enough microbial stimulus to keep it occupied and under control. It is by no means proven conclusively that excessive cleanliness, especially during the childhood years, is responsible for the rise in the incidence of conditions like asthma, but do we really want to do the experiment? "Go out and play in the dirt!" should be the admonition on the lips of parents everywhere, not "Don't forget to wash with the antibacterial soap!"

An amusing story about "antibacterial" products stars the antibacterial compound Triclosan. During the late 1990s, when antibacterial plastic cutting boards and plastic toys suddenly appeared in stores all over the United States, consumers doubtless assumed that something new had been added to these products. But this was not the case. Triclosan had been added to plastic products for years as a retardant of bacterial activities that can make plastics brittle and thus reduce their shelf life. All that changed between the cutting board that had no health claim and the one that was "antibacterial" was the label. Some advertising genius had simply decided to take advantage of the public's anxiety about "germs" by pointing to a compound that was already in the product.

Resistance to Antiseptics and Disinfectants

Resistance to antiseptics and disinfectants is still poorly understood, but it does occur. This is somewhat surprising because, unlike antibiotics, antiseptics and disinfectants have multiple targets. It would be impossible for a microbe to change all of its susceptible proteins and its DNA to become resistant, so target modification as a mechanism of resistance is out of the question. Given this, it is not surprising that the known antiseptic and disinfectant resistance mechanisms of bacteria seem to be designed to keep the antiseptic or disinfectant from reaching its target. Many antiseptics and disinfectants, especially those that attack membranes (e.g., QACs), are less effective against gram-negative bacteria than gram-positive bacteria. The reason may be that lipopolysaccharide (LPS) in the outer membrane of gram-negative bacteria prevents hydrophobic molecules from inserting themselves into the gram-negative outer membrane, while porins in the outer membrane help by restricting access of the antiseptics and disinfectants to the cytoplasmic membrane by limiting diffusion into the periplasmic space.

Another resistance mechanism that bacteria use against QACs and heavy metals is a cytoplasmic membrane pump, composed of proteins, that pumps the disinfectant out of the cell cytoplasm. Why this would make the bacteria resistant to QACs, which are thought to act mainly by dissolving membranes, is still unclear. Whatever the explanation, these pumps are fairly effective in protecting the bacteria from QACs. Genes encoding QAC pumps have been found on plasmids as well as in the chromosome. At one time, colloidal silver was billed as an antibacterial compound to which resistance would not emerge. Unfortunately, bacteria once again made fools of scientists and marketers who made that claim, and some bacteria have become resistant to silver, showing that such a thing can happen. Efflux pumps are the main mechanism of resistance.

The discovery that resistance to antiseptics and disinfectants can develop is disturbing because disinfectants and antiseptics are a vital line of defense against microbial infections. The fact that such resistance genes can also be linked genetically to antibiotic resistance genes on transmissible elements or integrons is even more disturbing because it means that use of disinfectants and antiseptics might actually select for maintenance of genes that confer resistance to antibiotics.

Triclosan once again enters the resistance story in an interesting way. Just as the textbooks had convinced us that antiseptics and disinfectants have relatively nonspecific targets compared to antibiotics, we expected that modes of resistance to disinfectants and antiseptics would tend to be similarly nonspecific, as exemplified by the efflux pumps that pump the compounds out of the cell. Triclosan has proved to be the first exception to that general rule. Although its action is as general as that of any other antiseptic or disinfectant, resistance to triclosan involves a specific enzyme. Apparently, triclosan inhibits an enzyme in a pathway for lipid biosynthesis. Resistance to triclosan can occur if mutations prevent triclosan from binding this enzyme. There are also efflux pumps that eject triclosan from the cell.

There seem to be two disinfectants to which bacteria may not be able to become resistant: alcohol and good old household bleach. It may not be an accident that hypochlorous acid, the active component in bleach, is also an antibacterial compound used by human white cells to destroy bacteria. This argument is not without flaws, however. These same cells use reactive forms of oxygen such as peroxide, and we know that bacteria can become resistant to peroxide by producing an enzyme that converts peroxide to water.

New Uses of Antiseptics and Disinfectants

Protecting antiseptics and disinfectants from abuse is important to preserve these important compounds for cases in which they are critically needed. Already, antiseptics are being used as an important part of the strategy for combating methicillin-resistant *Staphylococcus aureus* strains. Also, just as progress in identifying new antibiotics is slowing, so is progress toward the development of new nonantibiotic antibacterial compounds slowing. The public should be cautioned that overuse and abuse of antiseptics and disinfectants could be very injurious to their children, who may grow up in a world where the effectiveness of key antiseptics and disinfectants has been diminished.

Protecting these antibacterial compounds is now even more critical because of two new applications of such compounds. Triclosan is being explored as a control agent for the protozoan that causes malaria. Since resistance of this microbe to antimalarial drugs is rapidly increasing, triclosan could well make an important addition to the fight against rampant malaria. Another application of compounds like phenols, silver, and triclosan is to use them to impregnate plastic implants and indwelling catheters. Bacterial biofilms that form on medical plastic devices are becoming a very serious medical problem. Since biofilms that form on these devices are seldom directly treatable with antibiotics, the device has to be removed. For a venous catheter, this is a simple procedure: pull the catheter out of the patient's body. However, in the case of plastic heart valves or other internal implants, an operation is required to remove the contaminated device. Then yet another operation is needed to reimplant a new one after antibiotics have been used to eliminate remaining bacteria.

The public seems to be awakening to the importance of protecting antibiotics by preventing their overuse. Unfortunately, awareness of the dangerous path we are treading as we overuse antiseptics and disinfectants has not reached the same level of public consciousness and concern.

Issues to Ponder

1. OK, so you are socially responsible and you will not press your physician (at least not very hard) for an antibiotic prescription the next time you have a cold or sore throat. What are you going to do about disinfectant use? The use of antiseptics and disinfectants is selecting for resistant bacteria just as surely as the use of antibiotics is selecting for resistance to antibiotics. So, are you going to continue with your disinfectant-containing soaps and mouthwashes? Where does your sense of safety intersect with your sense of public responsibility?

2. Put in a different way, what would happen if we lost disinfectants and antiseptics because of resistance? Forget the preantibiotic world that was once so much touted by reporters. What would a predisinfectant, preantiseptic world look like? Are you becoming seriously depressed yet? Are you depressed enough to exercise limitations on the use of disinfectants and antiseptics? Is dirt good or what?

11

Antiviral, Antifungal, and Antiprotozoal Compounds

Throughout this book, the point has been made repeatedly that antibiotics are antibacterial compounds that have no efficacy against viruses. With a few exceptions, antibacterial compounds are also ineffective against fungi and protozoa. The reader who has made it to this last chapter may be wondering what types of compounds are available for treatment of diseases caused by microbes that are not bacteria. This chapter gives a few examples of such compounds, including examples in which antibiotics are actually used to treat infections that are caused by protozoa rather than bacteria.

Structures of antiviral, antifungal, and antiprotozoal compounds mentioned in this chapter are available in appendix 1.

Antiviral Compounds

For bacteria, there are broad-spectrum antibiotics. That is, antibiotics exist that are effective against most types of disease-causing bacteria. For viruses, by contrast, broad-spectrum compounds are not available, because different viruses can have very different approaches to reproducing themselves inside mammalian cells. Thus, a compound that is effective against one type of virus is not necessarily effective against a different virus that has a different reproduction strategy.

Scientists who develop antiviral compounds face a daunting challenge. Viruses are not free-living microbes. They are the ultimate freeloaders. They inject their genomes into mammalian cells and then highjack mammalian cell biosynthetic machinery and use it to make more viruses (Fig. 11.1). Because of this, there are not nearly as many steps in the viral life cycle that can be targeted by antiviral compounds without harming the host cell as there are for free-living microbes such as bacteria, fungi, and protozoa.

Lifestyles of the Small and Vicious

Viruses bind to the surface of a mammalian cell and proceed to invade the cell (Fig. 11.1). During the early stages of this invasion process, the virus sheds its surface covering to free its genome, which is released into the cytoplasm of the human cell. In most cases, the genome and a few associated proteins (the nucleoprotein core) move to the nucleus. In the nucleus, the viral genome is copied many times. Also, proteins that comprise the viral surface layer are produced at high levels. Viral proteins brought in with the viral genome may take part in this process, but most of the action is carried out by enzymes and cell components of the human cell. Then, the newly replicated viral particles exit the cell.

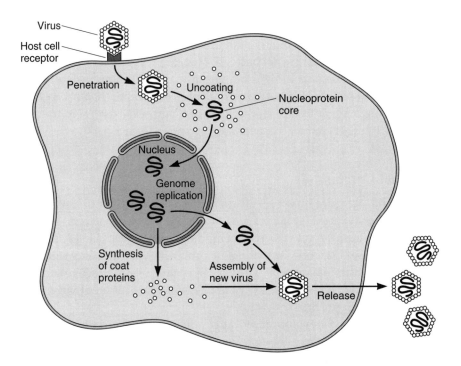

Figure 11.1   Generalized viral life cycle. The virus first binds to the surface of the host cell. It is then internalized. During this process the virus sheds its coat, thus releasing its genome. The genome then typically enters the nucleus, where it is copied. Viral proteins are synthesized in the cytoplasm. The newly synthesized proteins and viral genomes are assembled into new viral particles that are released from the cell.

Thus, there are only a few targets that can be hit by antiviral compounds: the viral proteins that mediate attachment to the human cell, the uncoating process, and viral proteins that participate in the copying of the viral genome. Perhaps the best strategy for stopping a virus in its tracks is preventing the virus from attaching to the human cell in the first place. This is the strategy used in vaccination against viral infections. Viruses have proteins on their surfaces that bind very specifically to some molecule on the host cell (viral receptor). This binding is a very tight lock-and-key type interaction. A vaccine containing viral surface proteins elicits antibodies that bind the viral surface proteins. Antibodies are complexes of blood proteins that are very bulky. A virus whose surface is coated with antibodies that bind viral proteins will not be able to attach itself to the receptor on the host cell.

Examples of Antiviral Compounds

There are also a number of antiviral compounds that can be administered directly to stop a viral infection. These compounds have the advantage that they do not have to be administered weeks before a person comes into contact with the virus, as is the case for vaccines. Some examples illustrate the different ways in which these compounds work.

Amantadine: an anti-influenza drug

After influenza virus attaches to the receptor it recognizes on the host cell surface (the nine-carbon sugar sialic acid), it stimulates the cell to take it up in a vesicle. The interior of the vesicle then becomes acidic due to action of human enzymes. The change in pH causes surface proteins in the coat of influenza virus to alter their conformation and allow the viral surface layer to be stripped off.

Amantadine stabilizes the viral proteins, preventing them from changing their configuration. Thus, even though the vesicle acidifies, the viral coat is not removed and the genome cannot be injected into the human cell cytoplasm. Amantadine is not 100% effective in eliminating influenza virus, but it can ameliorate the severity of a bout of influenza. Amantadine is generally reserved for elderly people with a flu infection, who are at particularly high risk for developing a secondary lung infection, bacterial pneumonia, which can be deadly.

Amantadine has been reserved for high risk patients because influenza viruses can become resistant to it. Influenza virus is one of the most rapidly mutating of all viruses, so resistance to amantadine can arise if the drug is used too widely.

Influenza virus has an RNA genome and as with all RNA viruses must ultimately make an RNA copy of its genome. So far, no anti-influenza drugs have been directed against the ability of influenza virus to make copies of its genome. Read on for an example of another virus with an RNA genome, HIV, where antiviral compounds are available that interfere with the viral reproduction of its genome.

Anti-HIV drugs

HIV, like influenza virus, has a genome that is composed of RNA rather than DNA. However, unlike the influenza virus, HIV does not simply make an RNA copy of its RNA genome; instead, it makes a DNA copy of its genome and then makes the RNA copies from that intermediate template.

The viral enzyme that makes the DNA copy of the HIV genome is called reverse transcriptase. This viral enzyme is the target of the anti-HIV drug AZT and of other nucleoside analogs. AZT looks like a nucleoside that would normally be incorporated by the reverse transcriptase into the growing DNA strand, but AZT lacks an important side chain that is necessary for adding the next base to the growing DNA chain. Thus, incorporating AZT causes synthesis of the DNA strand to be terminated.

The viral reverse transcriptase has a far higher affinity for AZT than human enzymes that synthesize DNA. This is the reason that AZT does not kill human cells, as would happen if AZT also terminated the synthesis of DNA in the human cells. Also, most human cells are terminally differentiated and no longer divide, so little human cell DNA synthesis occurs. AZT and similar inhibitors of reverse transcriptase are part of the highly active antiretroviral therapy (HAART) cocktail. Having more than one such inhibitor increases the efficacy of the therapy and may discourage the development of resistance.

Another component of the HAART cocktail is a protease inhibitor, which takes advantage of the way in which the proteins of HIV are made. Some HIV mRNAs contain genetic information for more than one protein. When the ribosomes translate such mRNAs, they make a polyprotein that consists of more than one protein fused in a single larger protein molecule. This polyprotein must be cleaved into individual viral proteins by a special viral protease. If this does not happen, the component viral proteins do not become active. Protease inhibitors inhibit the activity of the protease, thus preventing the polyprotein from being cleaved to produce active viral proteins.

Resistance to AZT and to protease inhibitors has already appeared. This happens when reverse transcriptase or the viral protease mutates enough that the drug no longer binds its target effectively.

Acyclovir: an anti-herpes virus drug

Cold sores caused by herpes simplex virus type I have been a fact of life for many people for many years. Then, in the 1980s, another herpes virus, herpes simplex virus type II, began to be a significant cause of sexually transmitted disease. HIV was definitely afoot by then but not yet widely covered in the media, and herpes virus type II hit the public consciousness as the first example of an incurable sexually transmitted disease—shades of syphilis in the 17th and 18th centuries, except that genital herpes was not as deadly as syphilis and AIDS. Still, it left its psychological mark. Infected people formed support groups and sought counseling. Today, it is difficult to appreciate how much of a psychological impact genital herpes had at the time when the public was first becoming aware of its existence.

There is now a drug that is quite effective against herpes simplex viruses: acyclovir. This drug does not, however, immediately cure herpes infections. The reason is the intriguing lifestyle of the virus. Herpes viruses replicate actively in epithelial cells, i.e., the cells that comprise mucous membranes of the mouth and genital tract. They also replicate in skin cells. The damage they do while they are actively reproducing themselves is evident from the cold sores or genital lesions they produce, yet after an attack of cold sores or genital lesions, the lesions subside and all seems to be well—until the lesions recur.

The recurrence phenomenon can be explained by the fact that herpes viruses do not simply replicate in epithelial cells but can also infect nerve cells. In nerve cells, however, they become quiescent, as if they had gone into hibernation. Thus, herpes viruses actively replicate in epithelial cells, and then when their human victim's immune response starts to kick in and destroy cells containing actively growing viruses, the viruses migrate to nearby neural ganglia and, in effect, go underground. Later, some stimulus, which we now label as "stress," to the infected person causes the viruses to reenter epithelial cells and begin once again to reproduce themselves actively. The cold sores or genital lesions return.

Acyclovir is most effective when the virus is growing actively in epithelial cells. The reason is that acyclovir interferes with replication of the viral DNA genome. Before that happens, however, acyclovir must be activated by a viral protein called thymidine kinase (Fig. 11.2). Normally, this enzyme phosphorylates bases that will be incorporated into DNA and may be a viral strategy for preempting its host cell's enzymes that initiate the same process. Acyclovir looks like a compound that can be incorporated in DNA, but if it is phosphorylated so that it can actually be incorporated in DNA by the viral enzyme, DNA polymerase, it stops the polymerization process, and viral replication ceases. Clearly, this strategy does not work if the virus is quiescent in nerve cells and not actively reproducing its genome. Thus, acyclovir acts only on the actively replicating viruses in epithelial cells. Herpes viruses may have invented the strategy: better to fight and run away and live to fight another day. It's not heroic, but it's very effective.

Herpes simplex virus has a DNA genome and does not mutate as rapidly as viruses with RNA genomes, like influenza virus and HIV, but resistance to acyclovir can occur. All the virus has to do is mutate either the enzyme that activates acyclovir in the first place or the DNA polymer-

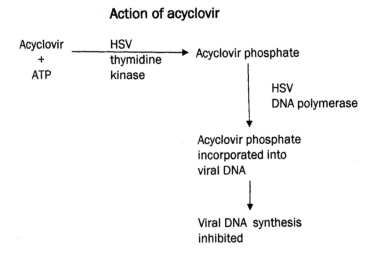

Figure 11.2   Mechanism of action of acyclovir. Acyclovir can be incorporated into viral DNA only if it is activated. The virus has an enzyme that activates acyclovir, but the human cell does not. Thus, activation occurs only when the virus is present in the cell.

ase that incorporates the phosphorylated acyclovir so that acyclovir is no longer effective in stopping viral replication.

Antifungal Compounds

Something approaching a broad-spectrum antifungal compound is possible because many fungi have their own form of cholesterol, called ergosterol. The enzymes involved in synthesis of ergosterol are different enough from the enzymes involved in the synthesis of cholesterol that antifungal compounds that target the fungal enzymes have little cross-reactivity with human cell cholesterol. However, many antifungal compounds nonetheless have a relatively narrow spectrum. We start with some of these narrow-spectrum compounds.

Compounds effective against skin and nail infections

A group of fungi called dermatophytes are responsible for such common infections as athlete's foot, jock itch, ringworm, and nail infections. An antifungal compound used to treat nail and skin infections is griseofulvin, also called grifulvin V (Appendix I). The mechanism of action of griseofulvin is not known for certain, but it is thought to inhibit the formation of microtubules. Microtubules are the protein complexes that bind to individual chromosomes during cell division (mitosis) and pull them apart as the cell divides.

Since human cells also have microtubules, you might expect that griseofulvin would be quite toxic for humans, but it actually has relatively few side effects. This is important, especially in the case of nail infections, because the duration of treatment for nail infections lasts many months. Griseofulvin is administered orally until the infected nail has been replaced by healthy nail. Due to the long-term administration and the fact that griseofulvin can cause liver damage in some people, the physician tests at intervals for liver enzymes that mark the onset of liver damage.

Another antifungal compound that is found in athlete's foot and jock itch sprays and powders is tolnaftate. This compound is an inhibitor of ergosterol synthesis.

Agents effective against candidiasis

Infections caused by *Candida albicans* (candidiasis) can range from minor skin infections (cradle cap) to vaginitis (vaginal itching and discharge) to life-threatening systemic infections in immunocompromised people. *C. albicans* is part of the human normal microbiota and is most commonly found in the mouth, throat, and vaginal tract. Some antifungal drugs that are used to treat candidiasis are shown in appendix 1. These include the azoles (e.g., diflucan, ketoconazole), the polyenes (e.g., amphotericin B, nystatin), and flucytosine. The azoles inhibit the synthesis of ergosterol. Reduced ergosterol synthesis destabilizes the fungal membrane and makes it difficult for the fungi to take up nutrients and perform other essential functions.

The polyenes bind ergosterol and form channels in the membrane. These compounds kill the fungus, which is a good feature, but they are also fairly toxic drugs. This toxicity probably occurs because they bind cholesterol, although not as tightly as ergosterol.

Flucytosine looks like a nucleotide (appendix 1) and when it is incorporated by the fungi into RNA, it disrupts transcription of the RNA. When the fungi mistakenly add a phosphate to flucytosine, they create a molecule that inhibits an enzyme called thymidylate synthase, which plays an important role in DNA synthesis. Thus, flucytosine, unlike most antimicrobial compounds, acts at more than one level.

As with tuberculosis therapy and some antiviral therapies, a combination of antifungal drugs is sometimes used to increase the effectiveness of the therapy and prevent the development of resistance. A commonly used combination is flucytosine and a polyene.

Agents effective against *Pneumocystis* pneumonia

Before effective therapy became available, a major killer of patients infected with HIV was a particularly lethal form of pneumonia caused by the fungus *Pneumocystis jiroveci* (formerly *P. carinii*—will they ever stop changing the names?). *P. jiroveci* is thought to be a common resident of the human mouth and upper respiratory tract, but it was not known to cause serious infections until severely immunocompromised patients such as AIDS patients came onto the clinical scene. *P. jiroveci* is unlike most other fungi in that it has cholesterol rather than ergosterol as its membrane sterol. Thus, drugs that target ergosterol are not effective.

A drug that is used widely to treat *P. jiroveci* pneumonia is pentamidine, a drug that inhibits DNA synthesis. The only other use of pentamidine is to treat infections caused by a group of protozoans called trypanosomes. An example of a trypanosomal disease is sleeping sickness. Antiprotozoal drugs do not normally work against fungi. *P. jiroveci* is also susceptible to the antibacterial agents sulfa-trimethoprim, so *P. jiroveci* is a real oddball in the antimicrobial susceptibility department.

Examples of Antiprotozoal Drugs

Antiprotozoal drugs, like antiviral drugs, tend to have a narrow spectrum. That is, the drug that is effective in treating one type of protozoal disease does not work on other types of protozoa. Just as the lack of broad-spectrum antiviral drugs reflects the diversity of viral replication strategies, so the lack of broad-spectrum antiprotozoal drugs reflects the diversity in the metabolism of protozoa. Oddly enough, although bacteria are far more diverse metabolically than protozoa, they nonetheless share certain traits such as their ribosomes which are different enough from human cell components to allow them to serve as targets for antibacterial compounds.

Antimalarial drugs

Malaria, a disease caused by a protozoan called *Plasmodium*, is one of the most common causes of death worldwide. Travelers from developed countries have to be careful to protect themselves against malaria when they travel to areas such as northern Africa and India, where the disease is once again out of control. Chloroquine, quinacrine, and primaquine are commonly used antimalarial drugs. Chloroquine and quinacrine (Atabrin) act by inserting themselves into the DNA double helix and interfering with DNA and RNA synthesis. Primaquine acts on mitochondria, the organelles of eukaryotic cells that generate energy for biosynthetic reaction, uptake of nutrients by the cell, and other essential cell processes.

How could *Plasmodium* species become resistant to drugs that act by inserting themselves nonspecifically into the DNA double helix? The protozoa cannot very well mutate their DNA drastically enough to prevent the drug from intercalating into their DNA, a very nonspecific drug action, so instead they simply expel the drug. To this end, the protozoa produce protein pumps located in their cytoplasmic membranes. These pumps pump the drugs out of the protozoal cell as fast as they enter it. By keeping the intracellular concentration of the drug too low to have an effect, they protect themselves from the drug. Given the widespread resistance of *Plasmodium* species to chloroquine and quinacrine, it is encouraging that recent breakthroughs seem to have been made in the development of a vaccine against malaria.

Protozoal diarrhea

People from developed countries often, do not consider diseases caused by protozoa to be a problem for them. After all, malaria and sleeping sickness are not seen in the United States and Europe, except in the occasional traveler. There are, however, two protozoal diseases that are widespread in the United States and have appeared in other developed countries: diarrhea caused by *Giardia intestinalis* (giardiasis) and diarrhea caused by *Cryptosporidium parvum*. *G. intestinalis* is considered by some to be one of the cutest microbial pathogens (Fig. 11.3), with its two nuclei that resemble large innocent-looking eyes, but it is one mean microbe. It causes a diarrhea that, unlike viral and bacterial diarrheas, does not subside within a few days but can hang on for months and be quite debilitating.

Giardiasis is spread by water contaminated with animal feces. Before you say, "Yuck, I would not drink such water," keep in mind that pristine-appearing mountain streams and even glacial lakes in the Rockies are a prime source of *G. intestinalis*. As with the protozoan itself, looks are deceiving. Hikers are becoming more sophisticated about the prevention of this disease, but *G. intestinalis* can also get into public water supplies that people have every right to assume are safe.

C. parvum is more pervasive, and outbreaks of disease caused by this protozoan have occurred in many parts of the United States, especially New York and the upper Midwest. One particularly memorable outbreak occurred in Milwaukee, Wis., when an error at a water treatment plant allowed water contaminated with the protozoan to be pumped into homes through the area.

Treatment of giardiasis provides another example of cases in which antibacterial compounds work against protozoa. Paromomycin (Hu-

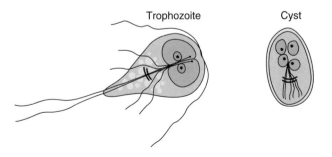

Figure 11.3   The two forms of *Giardia intestinalis*. The trophozoite is the actively replicating form found in the intestine. The cyst is the form that survives in the external environment.

matin) is an antibiotic that was first identified as a compound that inhibits bacterial protein synthesis, yet its main use today is to treat diarrhea caused by *G. intestinalis*. Paromomycin binds to the small subunit of the ribosome and causes the ribosome to make mistakes when it translates a protein.

The antibiotic metronidazole is also used to treat *G. intestinalis* infections and infections caused by some other protozoa. Metronidazole is an antibiotic that is most effective against microbes that grow best in the absence of oxygen, because these microbes have enzymes that convert metronidazole into a form that causes breaks in DNA. This feature of the drug caused initial concern about the possibility that it might be carcinogenic, but since human cells do not activate metronidazole to the form that causes breakage of DNA, this proved not to be a side effect of the antibiotic.

High-level resistance to metronidazole is rare in bacteria and protozoa, but intermediate resistance has been seen. The main mechanism of resistance is mutations in the enzymes that activate metronidazole to carry out that reaction. Obviously, complete loss of such enzymes would reduce the ability of the microbe to activate metronidazole, but such enzymes are essential for survival. This may be why intermediate resistance seems to predominate and high-level resistance is rare.

Antitrypanosomal drugs

Diseases caused by members of the genus *Trypanosoma* (trypanosomes) are quite uncommon in the United States and Europe but are all too common in the developing world. Two examples of trypanosomal diseases are sleeping sickness and leishmaniasis. The former is still common in Africa, and the latter is seen in many parts of South America. Both are spread by insects. Both are also very serious diseases that can be fatal, especially in infants and children. Two example of antitrypanosomal drugs are pentamidine and suramin. Pentamidine takes advantage of a peculiarity of trypanosomes. In addition to their mitochondria, trypanosomes have a modified mitochondrion called a kinetoplast. Pentamidine interferes somehow with the function of the kinetoplast. Given the fact that only trypanosomes have kinetoplasts, it is clear why pentamidine is a trypanosome-specific drug.

Suramin inhibits enzymes in organelles called glycosomes, which contain the enzymes that break down glucose. Many organisms, from microbes to mammals, have such enzymes, but they are not concentrated in a glycosome. For some reason, suramin inhibits the function of these enzymes in the glycosomes of trypanosomes much more effectively than in other cells. Probably, there is some feature of the glycosomes of trypanosomes that allows suramin to interfere with their function.

Issues to Ponder

1. Antiviral compounds rarely, if ever, cure a viral infection. Instead, they reduce the number of viruses below the threshold where disease symptoms begin to appear. Other antimicrobial compounds are able to clear their microbial target from the patient's body, i.e., to effect a complete cure. What might the explanation of this phenomenon be? What are the economic consequences of this property of antiviral compounds? What is likely to be the effect on viral development of resistance to the antiviral compound that is taken for years?

2. A theme that you have seen developing, starting with the case of the anti-tuberculosis drugs and continuing in the case of the antifungal and antiprotozoal drugs, is how little is known about how these drugs work, much less how the microbes become resistant to them. A great deal of effort has been put into understanding the mechanisms of action and resistance to antibiotics and to antiviral drugs, but this same level of attention has not been paid to the antifungal and antiprotozoal drugs. The rationalization for this uneven research coverage is that charity begins at home. We pay attention first to those diseases that affect us. This attitude is understandable, given that research funds are limited, but is it wise? Why might it be a good idea to pay more attention to the antifungal and antiprotozoal drugs, their mechanisms of action, and their mechanisms of resistance?

3. Most of us would agree that prevention is preferable to intervention when it comes to infectious diseases. In the case of malaria, the disease had nearly been brought under control in the 1970s, when a small problem arose. The "magic bullet" that was eliminating the mosquitoes that carry malaria was DDT. DDT was cheap and effective. Even when mosquitoes began to become resistant to its effects, DDT seemed to keep them away because the mosquitoes did not like its smell. Houses sprayed inside with DDT posed no general health threat to the environment or to the people who occupied the houses, but kept mosquitoes out. Then, in the West,

DDT became a dirty word. It now appears that much of the "scientific" evidence that DDT was a threat to animals and birds (it was never proven to be a threat to humans) was junk science. Nonetheless, the widespread bad reputation attached to DDT had the effect of making officials in developing countries worry about the safety of the insecticide and making manufacturers stop manufacturing it for fear of lawsuits.

The consequence of this today is that the malaria death rate, mostly among children, is back up to where it was before DDT spraying. Even if the people in the malaria belt could afford the cheaper drugs used to treat malaria, they are still in trouble because resistance to chloroquine has become widespread. Drugs that are still effective, like primaquine, are far too expensive and have side effects. Should the judicious use of DDT in malaria belt countries be reevaluated? Should those of us who live in developed countries mount a pro-DDT campaign?

Appendix 1

Structures of Antimicrobial Agents Mentioned in the Text

Antibacterial Agents (Antibiotics)

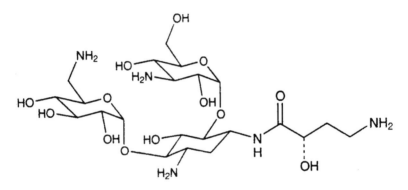

Amikacin

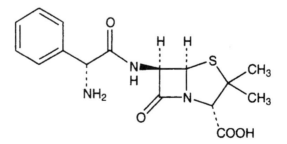

Ampicillin

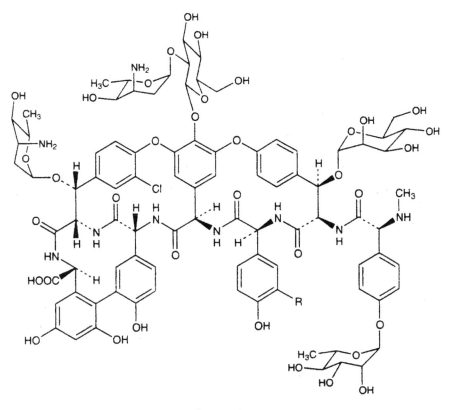

Avoparcin

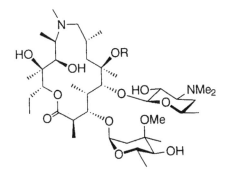

Azithromycin

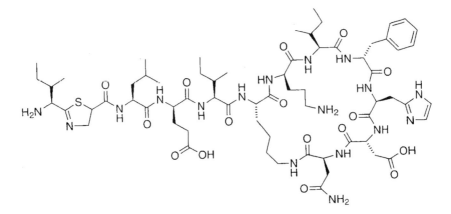

Bacitracin

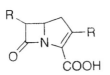

Carbapenem

Cephalosporin

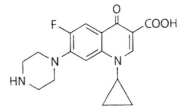

Ciprofloxacin

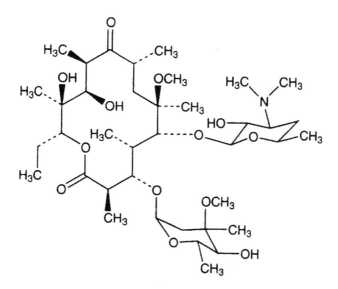

Clarithromycin

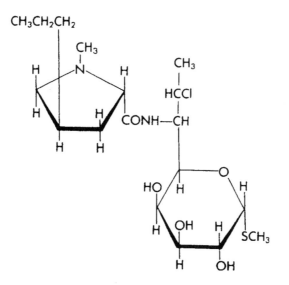

Clindamycin

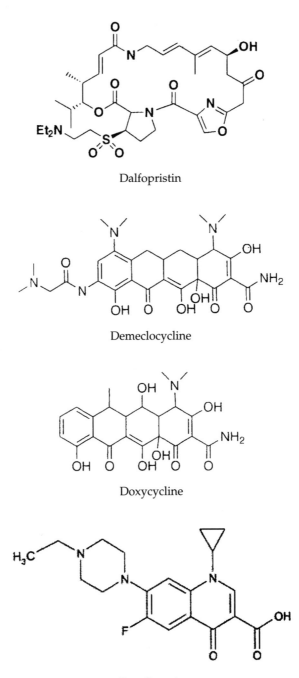

Dalfopristin

Demeclocycline

Doxycycline

Enrofloxacin

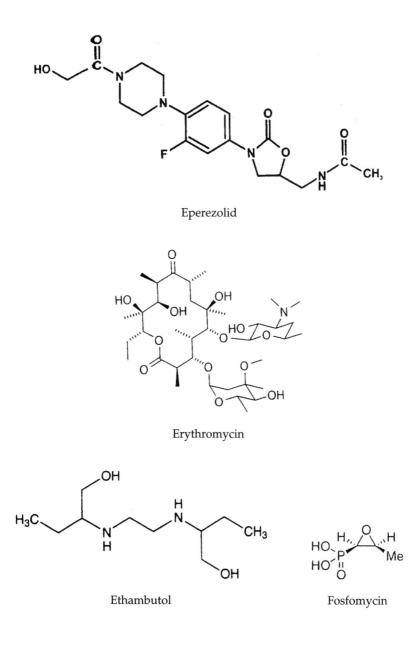

Eperezolid

Erythromycin

Ethambutol

Fosfomycin

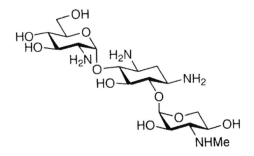

Gentamicin

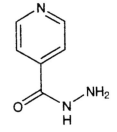

Isoniazid

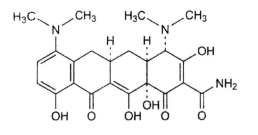

Kanamycin

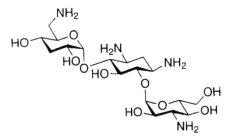

Linezolid

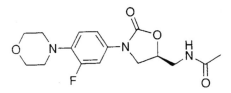

Minocycline

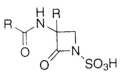

Monobactam

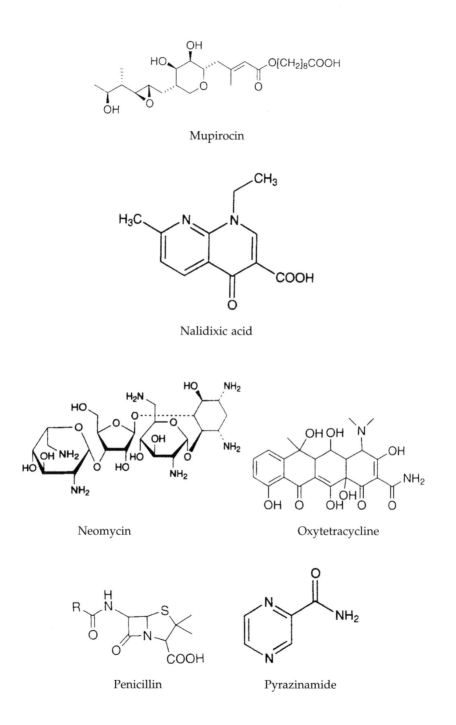

Mupirocin

Nalidixic acid

Neomycin

Oxytetracycline

Penicillin

Pyrazinamide

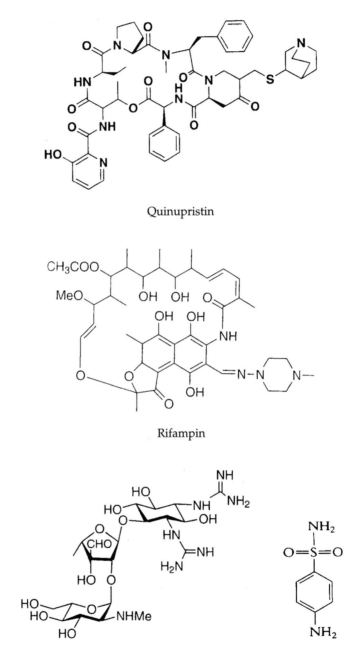

Quinupristin

Rifampin

Streptomycin   Sulfanilamide

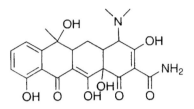

Tetracycline

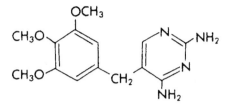

Trimethoprim

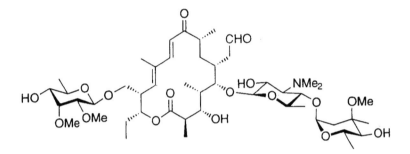

Tylosin

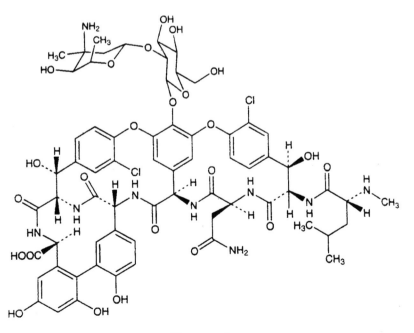

Vancomycin

Antiviral Agents

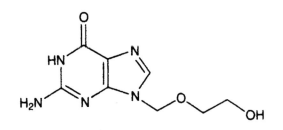

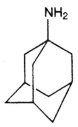

Acyclovir   Amantadine

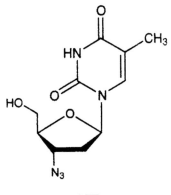

AZT

Protease inhibitors

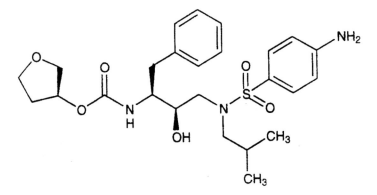

Amprenavir

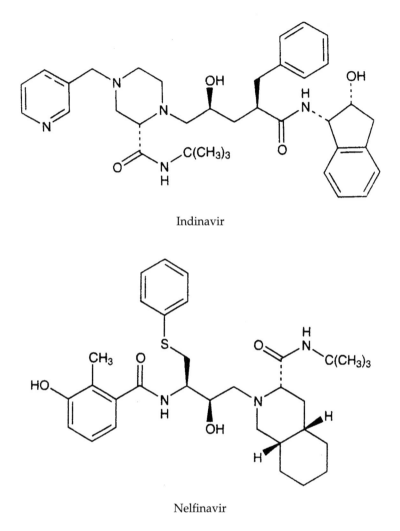

Indinavir

Nelfinavir

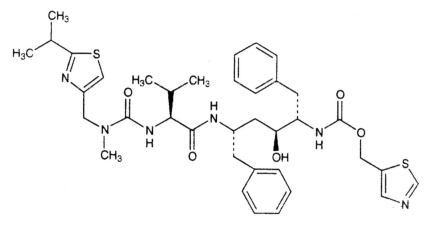

Ritonavir

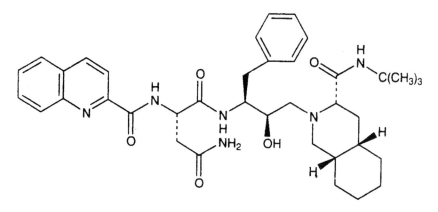

Saquinavir

Antifungal Agents

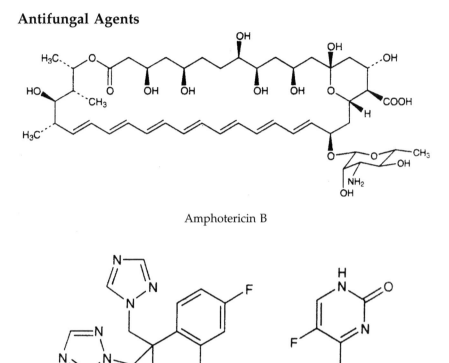

Amphotericin B

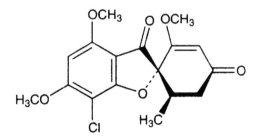

Diflucan   Flucytosine

Griseofulvin

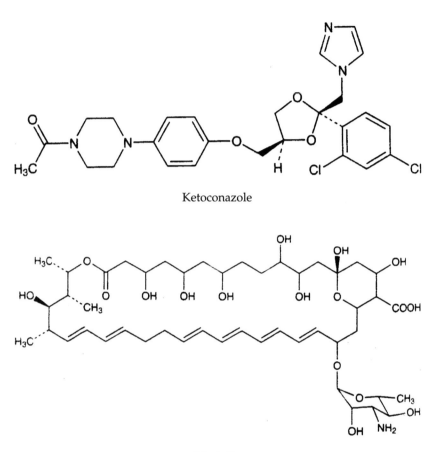

Ketoconazole

Nystatin

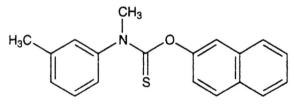

Tolnaftate

Agents Effective against Protozoal Parasites

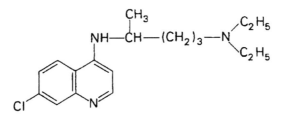

Chloroquine

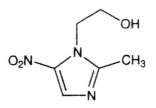

Metronidazole

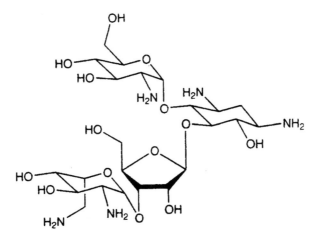

Paromomycin

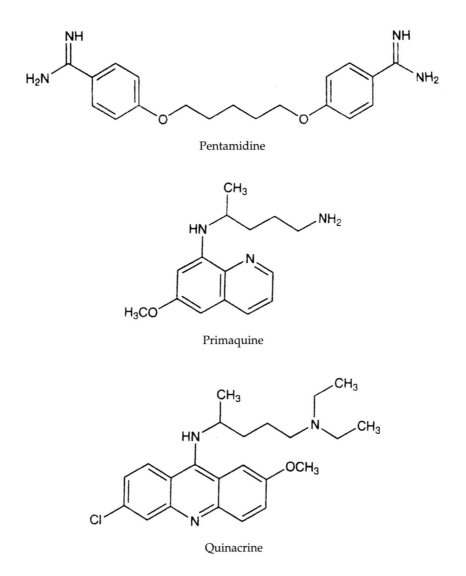

Pentamidine

Primaquine

Quinacrine

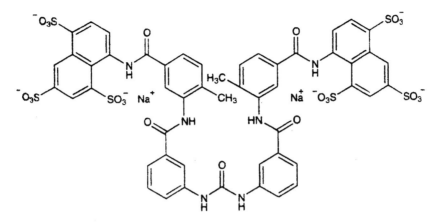

Suramin

Antiseptics and Disinfectants

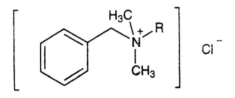

R = C_8H_{17} to $C_{18}H_{37}$

Benzalkonium chloride
(quaternary ammonium compound, or QAC)

Ethanol

Formaldehyde

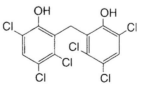

Hexachlorophene

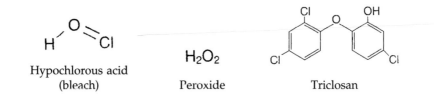

Hypochlorous acid
(bleach)

Peroxide

Triclosan

Appendix 2

How Clinical Laboratories Measure Resistance

A patient seeks help from a physician for a urinary tract infection or lies helpless in a hospital, ravaged by a postsurgical infection. The physician is tolerably certain that the infection is caused by a bacterium but is not sure which one is the culprit. In both cases, the physician's first move is to guess what antibiotic will be effective. In most hospitals, the physician will be guided by an antibiogram (Fig. A2.1).

There is a second line of inquiry that a good physician will take—assuming that his or her HMO has not discouraged this action—and that is to send a specimen to the laboratory for identification of the bacteria and for antibiotic susceptibility testing. The sequence of steps in analysis of a urine sample from a suspected urinary tract infection case is shown in Fig. A2.2. The laboratory will do a rapid test to determine whether the sample contains bacteria and the inflammatory (white) cells that indicate an infection is underway. This only takes minutes. The next steps take days. The bacterium in the specimen is first cultivated so that there are enough bacteria to test for antibiotic susceptibility. Usually this takes at least a day. Once the bacterium is isolated, it is inoculated into an automated test plate that contains different concentrations of a variety of antibiotics.

The antibiotics to be tested are chosen by the companies that produce these tests with a focus on antibiotics that are most likely to be effective for this type of infection. That is, a separate bank of antibiotics would be tested if the infection were a lung infection versus a wound infection. Susceptibility testing takes another day or two. If the patient has something less amenable to rapid cultivation of the causative microbe, such as tuberculosis, the cultivation and susceptibility testing could take over a month to yield results.

GRAM NEGATIVE BACTERIA
% susceptible

	Escherichia coli	Klebsiella pneumoniae	Proteus mirabilis	Pseudomonas aeruginosa
Ampicillin	70	0	87	Not done*
Cefazolin	91	89	97	Not done
Gentamicin	97	100	100	84
Trimethoprim/ Sulfamethoxazole	84	96	93	Not done
Ciprofloxacin	Not done	Not done	Not done	78

GRAM POSITIVE BACTERIA
% susceptible

	Staphylococcus aureus	Staphylococcus epidermidis	Other Staphylococci	Enterococcus faecalis
Ampicillin	Not done	Not done	Not done	93
Cephalothin	64	20	28	Not done
Clindamycin	66	Not done	Not done	Not done
Trimethoprim/ Sulfamethoxazole	99	67	54	Not done
Vancomycin	100	100	100	93

	Streptococcus pneumoniae	
	% susceptible	% intermediate
Penicillin	59	15
Cefotaxime	74	20
Erythromycin	57	--
Clindamycin	82	--
Trimethoprim/ Sulfamethoxazole	54	--
Vancomycin	100	--

*Not done – the bacterial species is known to generally be resistant to the antibiotic, so no test is done

Figure A2.1   Antibiogram similar to those provided in a hospital setting. The antibiogram tells the physician what percentage of isolates from that hospital that have been tested recently are resistant to a specific antibiotic. The antibiogram is revisited from time to time to reflect changes in resistance patterns.

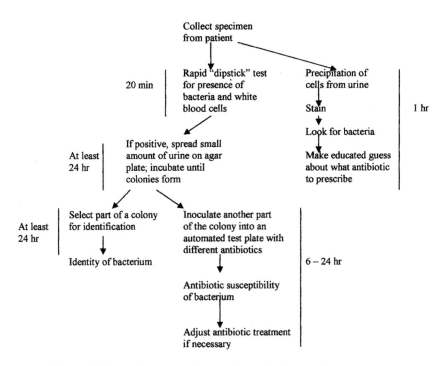

Figure A2.2   Sequence of steps for analyzing a urine specimen.

Widely Used Susceptibility Tests: the Disk Diffusion Test and the Microbroth Dilution Minimal Inhibitory Concentration (MIC) Test

The two types of test used most commonly to determine the antibiotic susceptibility of bacterial isolates are the disk diffusion test and the quantitative microbroth dilution MIC test. One type of disk diffusion test, the Kirby-Bauer test, is illustrated in Fig. A2.3. The bacterial strain to be tested is spread evenly over an agar plate. Then, disks containing different antibiotics are placed on the plate. The plate is incubated to allow the bacteria to grow wherever the concentration of antibiotic is not inhibitory. During the incubation period, antibiotics diffuse outward from the disks to form a concentration gradient. The effectiveness of the antibiotic is indicated by the diameter of the zone of no growth around the disk containing that antibiotic. Since different antibiotics diffuse outward from the disk at different rates and since the achievable levels of different antibiotics in blood differ, the same sized no-growth zone can indicate susceptibility

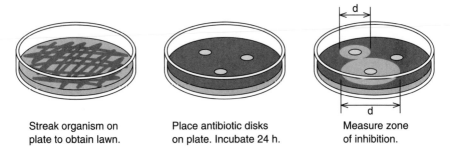

Streak organism on   Place antibiotic disks   Measure zone
plate to obtain lawn.   on plate. Incubate 24 h.   of inhibition.

Figure A2.3   The Kirby-Bauer test procedure. Bacteria are spread on the
surface of an agar plate to form a lawn. Disks containing known amounts
of antibiotic are placed on the lawn. The plate is incubated, and the diameters
of areas showing no growth are measured.

in one case and resistance in another. There are tables linking the zone
diameters to "susceptible," "intermediate," or "resistant" categories. *Susceptible* means ordinary dosages of that antibiotic will give drug levels in
blood and tissue high enough to kill or inhibit the growth of the bacterial
strain in the body. *Resistant* means normal dosages of the antibiotic will
not work. *Intermediate* means an elevated dosage might work.

A second type of test, the microbroth dilution MIC test is illustrated
in Fig. A2.4. Different wells of a microtiter plate contain successive twofold
increments of an antibiotic. The same numbers of bacteria are inoculated
into each well, and the plate is incubated to allow the bacteria to grow.
The lowest antibiotic concentration that prevents bacterial growth is called
the MIC. Some antibiotics merely inhibit the growth of the bacterium
without killing it. Such antibiotics are useful in people with a healthy
immune system because they can prevent bacterial growth long enough
to let the immune system clear the bacteria from the body. MICs can also
be reported as susceptible, intermediate, and resistant. As a rule of thumb,
the achievable levels of an antibiotic in blood and tissue should be 4 to 5
times higher than the MIC for the bacterial strain to be called susceptible.
This is because the body is constantly eliminating the antibiotic, and in
order to sustain a level of antibiotic in the blood or tissue at or above the
MIC for any length of time, the initial concentration must be severalfold
higher than the MIC. Currently, many clinical laboratories report MIC
results as susceptible, intermediate, or resistant rather than providing the
raw MIC scores, because if raw MIC scores are given, many physicians

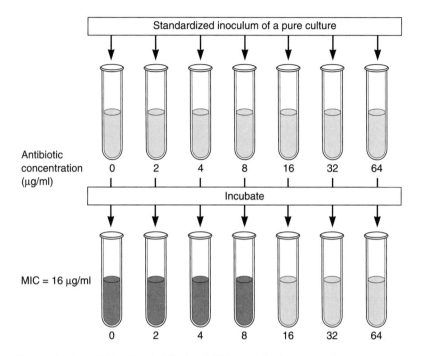

Figure A2.4   Microbroth dilution MIC assay. In this assay, the same number of bacteria is inoculated into each of a series of tubes containing increasing concentrations of an antibiotic. The tubes are checked for growth after incubating for 24 h. The first tube in which there is no detectable growth represents the MIC. In this figure, the MIC is 16 µg/ml.

choose the antibiotic with the lowest MIC. This is not necessarily the best choice, because the antibiotic with the lowest MIC might be more toxic than one with a higher MIC, and the achievable levels in blood might be lower. It is the relationship between achievable levels in blood and MIC that is important, not the absolute level of the MIC.

Microbroth dilution MIC tests have usually been done by inoculating a defined number of bacteria into the wells of a microtiter plate and then incubating the plate for a prescribed amount of time (usually 16 to 24 h) before reading the results. A testing procedure has recently been developed that speeds up this process considerably. The microtiter plates are placed in a machine that measures the turbidity in the well (an indication of microbial growth) at intervals over a 4- to 5-h period. Differences in growth rate can be used to determine the MIC. Thus, results can be obtained within about 6 h after the microbe is isolated from the specimen.

In patients with severely compromised immune systems, an antibiotic that kills bacteria is preferred over one that merely inhibits their growth. Accordingly, a second type of test, called the minimum bactericidal concentration (MBC) test, is sometimes done. A sample from each well of an MIC test that exhibits no growth is streaked onto an agar plate that does not contain any antibiotic and is incubated to allow viable bacteria to grow. The lowest antibiotic concentration that yields no bacterial growth on the agar plate is called the MBC. MBC testing is somewhat controversial. For one thing, the MBC procedure has not undergone the same sort of rigorous standardization procedures that have been used to validate the disk diffusion and microbroth dilution MIC tests. For another, MBC tests done with laboratory medium may not be predictive of actual antimicrobial performance in patients. However, despite general dissatisfaction with the current version of the MBC test, the increasing number of patients whose underlying conditions require bactericidal rather than bacteriostatic antibiotics keeps interest in this type of test alive.

The microbroth dilution test is more quantitative than the disk diffusion test, but it is more expensive to run. A third type of susceptibility test, the E-test, is a more quantitative variant of the disk diffusion test and is considered to give the most accurate MIC. In this test, plastic strips containing a gradient of an antimicrobial agent are placed on an agar plate spread with bacteria. The strips are imprinted with a scale showing the concentration of the antibiotic. The MIC is the value on the strip where the elliptical zone of inhibition intersects the strip.

It is usually necessary to begin therapy before results of susceptibility tests are known. By considering the site of infection, the patient's symptoms, and underlying conditions, the physician can usually narrow the diagnosis to one or several microbes. Help in making an educated guess about what therapeutic agent is most likely to be effective comes from past information about the susceptibility pattern of the suspected agents in the physician's hospital or community. This information is provided on an antibiogram (Fig. A2.1).

Issue to Ponder

Why are clinical laboratories still using the same old tests? We don't know the answer to this question. The problem is that neither physician nor patient can wait days or weeks for results to guide treatment, especially in this era of antibiotic resistance. Thus, physicians act first on their best guess as to effective therapy. If you are wondering why modern technology has not speeded up the process of diagnosis of an infection and identification of the most effective therapy, you are not alone. There are new DNA-based tests that utilize polymerase chain reaction—a form of amplification of specific DNA sequences—that could give answers within hours, yet these are only used in a small minority of diagnoses. For once, the problem is not cost. These procedures are relatively cheap as well as rapid.

Selected Reading
Louie, M., and F. R. Cockerill III. 2001. Susceptibility testing. Phenotypic and genotypic tests for bacteria and mycobacteria. *Infect. Dis. Clin. N. Am.* **15:**1205–1226.

Suggested Reading

Aiello, A. E., and E. Larson. 2003. Antibacterial cleaning and hygiene products as an emerging risk factor for antibiotic resistance in the community. *Lancet Infect. Dis.* **3:**501–506.

Allen, P. L. 2000. *The Wages of Sin: Sex and Disease, Past and Present.* University of Chicago Press, Chicago, Ill.

Andremont, A. 2003. Commensal flora may play key role in spreading antibiotic resistance. *ASM News* **69:**601–607.

Aymes, S. G. B. 2001. *Magic Bullets, Lost Horizons: the Rise and Fall of Antibiotics.* Taylor & Francis Books, New York, N.Y.

Bailar, J. C., III, and K. Travers. 2002. Review of assessments of the human health risk associated with the use of antimicrobial agents in agriculture. *Clin. Infect. Dis.* **34**(Suppl. 3):S135–S143.

Boyce, J. M. 2001. Antiseptic technology: access, affordability, and acceptance. *Emerg. Infect. Dis.* **7:**231–233.

Callahan, G. N. 2003. Eating dirt. *Emerg. Infect. Dis.* **9:**1016–1021.

Centers for Disease Control and Prevention. 2002. *Staphylococcus aureus* resistant to vancomycin—United States, 2002. *Morb. Mortal. Wkly. Rep.* **51:**565–567.

Cohen, M. L. 1992. Epidemiology of drug resistance: implications for a post-antimicrobial era. *Science* **257:**1050–1055.

Davies, J. 1996. Bacteria on the rampage. *Nature* **383:**219–220.

Ferber, D. 2003. WHO advises kicking the livestock antibiotic habit. *Science* **301:**1027.

Ferber, D. 2003. Triple-threat microbe gained powers from another bug. *Science* **302:**1488.

Goozner, M. 2004. *The $800 Million Pill.* University of California Press, Berkeley, Calif.

Heinemann, J. A., R. G. Ankenbauer, and C. F. Amabile-Cuevas. 2000. Do antibiotics maintain antibiotic resistance? *Drug Disc. Today* **5:** 195–204

James, C. W., and C. Gurk-Turner. 2001. Cross-reactivity of beta-lactam antibiotics. *Baylor Univ. Med. Center Proc.* **14:**106–107.

Jones, J. 1993. *Bad Blood. The Tuskegee Syphilis Experiment.* Simon and Schuster, New York, N.Y.

Kiester, E., Jr. 1990. A curiosity turned into the first silver bullet against death. *Smithsonian* **20**(8):172–187.

Klein, G., C. Hallmann, I. A. Casas, J. Abad, J. Louwers, and G. Reuter. 2000. Exclusion of *vanA*, *vanB* and *vanC* type glycopeptide resistance in strains of *Lactobacillus reuteri* and *Lactobacillus rhamnosus* used as probiotics by polymerase chain reaction and hybridization methods. *J. Appl. Microbiol.* **89:**815–824.

Koch, A. L. 2003. Bacterial wall as target for attack: past, present, and future research. *Clin. Microbiol. Rev.* **16:**673–687.

Kress, N. 1995. Evolution. *Isaac Asimov's Sci. Fiction Magazine,* October.

Levy, S. B. 1998. The challenge of antibiotic resistance. *Sci. Am.* **275:**46–53.

Levy, S. B. 2002. *The Antibiotic Paradox: How the Misuse of Antibiotics Destroys Their Curative Powers,* 2nd ed. Perseus Publishing, Cambridge, Mass.

Maki, D. G., and P. A. Tambyah. 2001. Engineering out the risk of infection with urinary catheters. *Emerg. Infect. Dis.* **7:**1–6.

Mann, T. 1924. *The Magic Mountain.* Knopf Publishing Co., New York, N.Y. (1927 English translation).

McDonnell, G., and D. Russell. 1999. Antiseptics and disinfectants: activity, action and resistance. *Clin. Microbiol. Rev.* **12:**147–179.

McEwen, S. A., and P. J. Fedorka-Cray. 2002. Antimicrobial use and resistance in animals. *Clin. Infect. Dis.* **34**(Suppl. 3):S93–S106.

McGowan, J. E., Jr. 2001. Economic impact of antimicrobial resistance. *Emerg. Infect. Dis.* **7:**286–292.

McMurry, L. M., M. Oethinger, and S. B. Levy. 1998. Triclosan targets lipid synthesis. *Nature* **394:**531–532.

Moberg, C., and Z. Cohn (ed.). 1990. *Launching the Antibiotic Era.* Rockefeller University Press, New York, N.Y.

Noble, W. C., Z. Virani, and R. G. A. Cree. 1992. Co-transfer of vancomycin and other resistance genes from *Enterococcus faecalis* NCTC 12202 to *Staphylococcus aureus*. *FEMS Microbiol. Lett.* **93:**195–198.

O'Brien, T. F. 2002. Emergence, spread, and environmental effect of antimicrobial resistance: how use of an antimicrobial anywhere can increase resistance to any antimicrobial anywhere else. *Clin. Infect. Dis.* **34**(Suppl. 3):S78–S84.

Parascandola, J. (ed.). 1980. *The History of Antibiotics: a Symposium.* American Institute of the History of Pharmacy, Madison, Wis.

Pittet, D. 2001. Improving adherence to hand hygiene practice: a multidisciplinary approach. *Emerg. Infect. Dis.* **7**:234–240.

Projan, S. J. 2003. Why is big Pharma getting out of antibacterial drug discovery? *Curr. Opin. Microbiol.* **6**:1–4.

Rao, S. P., A. Surolia, and N. Surolia. 2003. Triclosan: a shot in the arm for antimalarial chemotherapy. *Mol. Cell. Biochem.* **253**:55–63.

Rutala, W. A., and D. J. Webe. 2001. New disinfection and sterilization methods. *Emerg. Infect. Dis.* **7**:348–353.

Salyers, A. A., A. Gupta, and Y. Wang. 2004. Human intestinal bacteria as reservoirs for antibiotic resistance genes. *Trends Microbiol.* **12**:412–416.

Schmidt, C. W. 2002. Antibiotic resistance. More at stake than steak. *Environ. Health Perspect.* **110**:A396–A402.

Vanden Eng, J., R. Marcus, J. L. Hadler, B. Imhoff, D. J. Vugia, P. R. Cieslak, E. Zell, V. Deneen, K. G. McCombs, S. M. Zansky, M. A. Hawkins, and R. E. Besser. 2003. Consumer attitudes and use of antibiotics. *Emerg. Infect. Dis.* **9**:1128–1135.

Walsh, C. 2003. *Antibiotics: Actions, Origins, Resistance.* ASM Press, Washington, D.C.

Wegener, H. C. 2003. Ending the use of antimicrobial growth promoters is making a difference. *ASM News* **69**:443–448

Wilson, J. F. 2002. Renewing the fight against bacteria. *Scientist* **16**(5): 22–23.

Index